Security Management
A Critical Thinking Approach

Occupational Safety and Health Guide Series

Series Editor

Thomas D. Schneid
Eastern Kentucky University
Richmond, Kentucky

Published Titles

The Comprehensive Handbook of School Safety, *E. Scott Dunlap*

Corporate Safety Compliance: OSHA, Ethics, and the Law, *Thomas D. Schneid*

Creative Safety Solutions, *Thomas D. Schneid*

Disaster Management and Preparedness, *Thomas D. Schneid and Larry R. Collins*

Discrimination Law Issues for the Safety Professional, *Thomas D. Schneid*

Labor and Employment Issues for the Safety Professional, *Thomas D. Schneid*

Loss Control Auditing: A Guide for Conducting Fire, Safety, and Security Audits, *E. Scott Dunlap*

Loss Prevention and Safety Control: Terms and Definitions, *Dennis P. Nolan*

Managing Workers' Compensation: A Guide to Injury Reduction and Effective Claim Management, *Keith R. Wertz and James J. Bryant*

Motor Carrier Safety: A Guide to Regulatory Compliance, *E. Scott Dunlap*

Occupational Health Guide to Violence in the Workplace, *Thomas D. Schneid*

Physical Hazards of the Workplace, *Larry R. Collins and Thomas D. Schneid*

Safety Performance in a Lean Environment: A Guide to Building Safety into a Process, *Paul F. English*

Security Management: A Critical Thinking Approach, *Michael Land, Truett Ricks, and Bobby Ricks*

Forthcoming Titles

Workplace Safety and Health: Assessing Current Practices and Promoting Change in the Profession, *Thomas D. Schneid*

Security Management
A Critical Thinking Approach

Michael Land
Truett Ricks
Bobby Ricks

CRC Press
Taylor & Francis Group
Boca Raton London New York

CRC Press is an imprint of the
Taylor & Francis Group, an **informa** business

CRC Press
Taylor & Francis Group
6000 Broken Sound Parkway NW, Suite 300
Boca Raton, FL 33487-2742

First issued in paperback 2019

ISBN-13: 978-1-4665-6177-9 (hbk)
ISBN-13: 978-0-367-37912-4 (pbk)

Library of Congress Cataloging-in-Publication Data

Land, Michael.
 Security management : a critical thinking approach / Michael Land, Truett Ricks, and Bobby Ricks.
 pages cm. -- (Occupational safety & health guide series ; 14)
 Includes bibliographical references and index.
 ISBN 978-1-4665-6177-9
 1. Business enterprises--Security measures. 2. Office buildings--Security measures. 3. Computer security. I. Ricks, Truett A. II. Ricks, Bob. III. Title.

HD61.5.L36 2013
658.4'7--dc23   2013008519

Contents

Contents

Preface

Security is a paradox. It is often viewed as intrusive, unwanted, a hassle, or something that limits our personal, if not professional, freedoms. While, at the same time, if we need security, we can never have enough. It is the intention of this book to provide security practitioners with the ability to critically examine their organizational environment and provide a mechanism to make it secure while assuming an optimal relationship between obtrusion and necessity.

Security management is a very diverse function for which security practitioners must plan, manage people, administer budgets, process information, as well as prepare for emergencies, violence, and other loss scenarios. This book is written for students who would like to become professional security managers, as well as for the security manager who wants to be a more effective administrator. *Security Management: A Critical Thinking Approach* provides a disciplined process to look at the security management functions to better tailor security to any organization.

We present the most accurately balanced picture of security functions by using a critical thinking approach to interpret data as a tool for providing more effective security management. It is not the intent of this book to attempt to instruct someone on how to secure an environment with technical proficiency or to merely impart facts pertaining to security management. Although there is commonality in many aspects of security and potential loss environments, all situations are going to be unique. The premise in this text is to create a practitioner who will completely examine the environment and make informed, well-thought-out judgments to tailor a security program to fit a specific organization.

The book will enable students to think clearly and critically about the process of security management. It emphasizes the ability to articulate the differing aspects of business and security management by reasoning through complex problems in the changing organizational landscape. We guide students through the stages of the critical thinking process to formulate a practical program for security management.

We also emphasize core security management competencies of planning, organizing, staffing, and leading while providing a process to critically analyze these functions. The book stresses the benefits of using a methodical critical thinking process in building a comprehensive safety management system, and specifically addresses information security, cyber security, energy-sector security, chemical security, and general security management, while utilizing a critical thinking framework.

Security Management: A Critical Thinking Approach goes farther than other available books regarding security management because it not only provides fundamental concepts in security, but also creates informed, critical, and creative security managers who communicate effectively in their environment.

Michael Land
Truett Ricks
Bobby Ricks

The Authors

Michael Land, PhD, is a student success specialist for Eastern Kentucky University, College of Justice & Safety (Richmond). In addition to his over 20 years at EKU, he has been a consultant for both state and federal agencies in different facets of safe/secure technology design and implementation. Dr. Land has taught a number of courses in security, occupational safety, and loss prevention throughout his career at EKU. He holds a BS in security and loss prevention and an MS in loss prevention administration, both from Eastern Kentucky University. Dr. Land received his doctorate in educational leadership from Lincoln Memorial University (Harrogate, Tennessee).

Truett A. Ricks, PhD, is a retired dean of the College of Law Enforcement (now the College of Justice & Safety) at Eastern Kentucky University. He was dean for 14 years. Prior to becoming Dean in the College of Justice & Safety, Dr. Ricks served as associate dean and professor at the university.

In 1975, Dr. Ricks took a two-year sabbatical from teaching at Eastern Kentucky University and was appointed commissioner of the Kentucky State Police. During his tenure as commissioner, he implemented the promotional procedure that was enacted and placed in Kentucky Revised Statutes (K.R.S.) Chapter 16 by the 1976 state general assembly. That procedure remains in place to this day.

Dr. Ricks began his career in law enforcement as a clerk with the Federal Bureau of Investigation (FBI) in Washington, D.C. While serving as a patrol officer and as detective sergeant with the Memphis Police Department (Tennessee), he received his BA and MBA at the University of Memphis. He earned his PhD at Florida State University (Tallahassee) where he also was employed as an assistant professor in the Department of Criminology.

He is the author of several books on security and police management, and has written numerous articles on the subject of public management and security. Dr. Ricks is president of Ricks International Consulting, Inc., and presently provides consulting services to clients in the United States, Great Britain, and Nigeria. Formerly, he was the owner of an electronic/physical security company in Richmond, Kentucky. He served as a member of the Kentucky Crime Commission and the Kentucky Law Enforcement Council, as well as serving as the chair for the Kentucky Law Enforcement Council and as chair of the curriculum committee.

Bobby Ricks, JD is a faculty member in homeland security at Midway College, Midway, Kentucky. He graduated from Eastern Kentucky University with a BS in police administration, and the University of Memphis with a JD degree. He earned the designation Certified Protection Professional (CPP) through ASIS International (1986).

His law enforcement career began as a law enforcement specialist with the U.S. Air Force Security Forces. After being honorably discharged from the Air Force,

Ricks worked as a patrolman and then director of Crime Prevention in the Richmond Police Department (Kentucky). Following law school, he worked as a special agent with the FBI. He then transferred to the Federal Law Enforcement Training Center in Glynco, Georgia, where he was a lead instructor, senior instructor, branch chief, and division chief of the Legal Division. Following his federal career, Ricks moved back to Kentucky where he was the attorney supervisor at the Kentucky Department of Criminal Justice Training.

Ricks was a member of the adjunct faculty at Georgia Military College, Valdosta State University, Coastal Georgia Community College, and Eastern Kentucky University. He has prepared distance-learning courses for Ashworth College and facilitated courses for Boston University. In addition to his law enforcement and legal coursework, Ricks has taught leadership, ethics, management and supervision, and quality management programs. He has written articles on law, law enforcement, and security for regional and national publications.

1 Introduction to Security Management

Mankind's entry into the twenty-first century was met with a blast of insecurity. The United States, with the cold war over, domestic terrorism at an all-time low, and a thriving economy, achieved a feeling of accomplishment and content. There were world problems, namely, struggles in the Middle East between Israel and Palestine, adjustment to a global economy, and the economic and ethnic struggles in the new governments formed after the dissolution of the Soviet Union. That was "over there." Al Qaeda and other terrorist groups had attacked U.S. people and property overseas: the USS Cole in Yemen, soldiers in Somalia, and U.S. Embassy bombings in Tanzania and Kenya. Again, that was "over there," and we felt safe at home in the continental United States. On September 11, 2001, that changed. The attacks on the World Trade Center in New York and the Pentagon in Washington, D.C., as well as the hijacked Flight 93 crash in Pennsylvania, signaled to the citizens in the United States that they were not safe. Overseas, attacks on Spain and the United Kingdom had the effect of creating unrest throughout Europe and the free world.

People want safety at home and in the workplace. The awareness of threats against our communities requires us to be proactive in identifying threats. The USA Patriot Act (2002) and the reorganization of government under the Department of Homeland Security (DHS) was an answer to that concern. Facility security, awareness programs, and intelligence gathering and sharing are components of the effort to secure the homeland.

Ignoring these dangers and risks is unacceptable. Being proactive directs us to look at our individual communities, identify its assets, recognize potential targets, and develop a strategy to prioritize and plan against events, whether manmade or natural.

Physical Security is directed toward prevention. Homeland Security Presidential Directives established policies to strengthen the preparedness through prevention and organized response to terrorist attacks, major disasters, and other emergencies. The Major Preparedness Goal is to establish measurable readiness targets that balance the threat and magnitude of terrorist attacks, major disasters, and other emergencies with the resources to prevent, respond to, and recover from such incidents (HSPD, 2011).

While DHS initiatives answer a need for government to be secure and for the government to identify potential threats, whether public or private, it does not fully answer the need of private industry in securing their facilities and operations. Private enterprise implements security measures as a means for maintaining a profit for the organization. They see terrorist threats as a loss to profit. In addition to a terrorist threat, losses through internal and external crimes, espionage, fire, and other hazards, whether natural or manmade, are of concern to businesses and industry.

Through risk assessment and management, private enterprise identifies hazards and weighs the risk of not providing a level of security against the associated cost should an event occur. For example, the decision to purchase an insurance policy to indemnify a hazard is weighed against the probability of the event occurring and the consequences of the event.

In addition to direct internal and external criminal threats, employee safety affects morale, which, in turn, impacts productivity. Unhappy customers who may attack employees or damage property or irate employees who retaliate against the organization cause others to feel frightened when they go to work. Employees with domestic issues may cause events to flow into the workplace causing fear and loss of productivity due to stress. Likewise, downtime from workplace evacuations due to bomb threats, false alarms, and other disruptions add to elevated stress and low morale.

As the government seeks to protect its citizens and property, there is a need for additional support from private security. Private enterprise spends billions of dollars on security hardware: locks, alarms, fencing, closed circuit television systems, and other equipment in providing for a safe workplace. Governmental laws or regulations, insurance requirements, and partnership agreements may cause an enterprise to maintain a level of security, which, in turn, fuels the need for security hardware, personnel, and training.

Security is about risk reduction. Whether a government agency or a private enterprise, physical security focuses on four main ideas: continuity of operations, asset retention, workplace safety, and mitigation of damages through physical loss, employee productivity, and lawsuits. When an event affects an operation, the time it takes for the organization to return to normal operations affects the ability to continue its mission. Loss of assets through internal or external theft, espionage, or waste causes loss of profit through replacement costs and increases the cost of goods or services to the consumer, or, even, higher taxes in government operations.

Having a workplace where employees feel safe provides them a comfort level that allows them to work at peak efficiency without fear. Providing a safe and secure facility for occupants leads to increased productivity, which lowers costs and raises profits. For example, an apartment or office complex must provide a safe environment to attract tenants, and a retail store must provide a safe environment to attract customers. If tenants or customers do not feel safe, they will take their business elsewhere. In addition to these benefits, the risk of loss through lawsuits is lessened. A facility that implements and maintains a level of security lowers the risk of lawsuits and the amount of a judgment in the event of a lawsuit.

This is where a security manager steps in. The security manager must be good at managing and leading. The leader/manager must possess a measure of technical knowledge, be able to work with other people, and be able to develop and conceptualize ideas. The degree of each skill varies with the position and role of the security manager. First-line supervisors work closely with the line staff and need more technical skill in their work activity, but less conceptual skill. Conversely, the executive would spend more time utilizing conceptual skills in moving the organization forward than he would his knowledge of technical operations. A midlevel manager would possess some technical and some conceptual skill in roughly equal amounts. The ability to work with others is a necessity at all levels of management.

Chart on Skills

We manage things; we lead people. Managing operations focuses on efficient processes and process control, while leaders focus on the people and creating the organizational culture.

Leadership is defined as "a process whereby an individual influences a group of individuals to achieve a common goal" (Northouse, 2006). What makes a good and effective leader is the ability to influence others. The leader sets the vision of the organization or group. The leader "sells" the vision to the group to inspire and motivate them to achieve specific goals. The leader communicates the goals to the group and makes clear the expectations in order to achieve the goals. Leaders seek commitment from the group to strive for the goals. They empower the group with the authority necessary to complete the tasks.

Managers coordinate assignments and provide the resources to allow their staff to perform their assignments efficiently and effectively. They provide structure and outline processes necessary to maximize productivity and to prevent waste.

Managers establish policy and rules to provide structure for defining work procedures.

Managers are problem solvers. They examine work processes and identify ways to eliminate unnecessary tasks and streamline the workflow.

HOW TO USE THIS BOOK

Management and leadership skills can be learned. This book will explore fundamental concepts of both leadership and management, and then provide applications of these concepts to the role of the security manager. This chapter is a synopsis of each chapter will lead you to key information.

Chapter 2: Explores the fundamentals of critical thinking and analysis. Managers must be ready to analyze problems and understand how information is influenced by bias and perceptions.

Chapter 3: Examines the core competencies of managers: planning, organizing, and supervising.

Chapter 4: Tells us how to build a security management system. Strategic planning for security goals: risk assessments and surveys for benchmarking

Chapter 5: Policies and procedures. Defines a policy and how it affects mission and vision, rules, general day-to-day activities, and allows flexibility for unforeseen circumstances.

Chapter 6: Staffing. Training, supervising, and motivating labor relations.

Chapter 7: Physical controls. An overview of the physical security process. Focuses on standards and benchmarks for security applications.

Chapter 8: Identifies leadership characteristics, integrity, confidence, and vision.

Chapter 9: Risk management. The basis to security management; why we need security versus the cost of not having security.

Chapter 10: Computer information security.

Chapter 11: Cyber security.

Chapter 12: Security investigations.

Chapter 13: Security in the energy sector.

Chapter 14: Security in the chemical sector.

Chapter 15: The impact of 9/11 on security management.

Law and liability, threat and risk assessment, emergency preparedness, and security design and crime prevention through environmental design form the beginning of identifying a need for security awareness. The basic questions of where we are, where do we need to be, and how do we get there identify security needs and form our objective for efficient and effective security programs. Risk benefit and cost analysis round out the formula for determining security needs.

Pulling it all together, the book examines policies and procedures necessary for an efficient and effective security program. For all of the security systems to work, the overall security policy must address issues of access control: employee, visitor, and package screening, and deliveries. Awareness programs must be developed to educate personnel on threats and security procedures. Finally, employee background checks and visitor access requirements are examined to round out the security program.

If you do the little jobs well, the big ones will tend to take care of themselves.

Dale Carnegie

Too many times, we look at the "big picture" and fail to see the little things. We are prepared for the terrorist attacks, but ignore incidents we see as minor distractions or nuisances. The broken window theory teaches us that small items, such as broken windows, graffiti, unkempt lawns, etc., invite small crimes like trespassing and criminal mischief. Such activity unheeded evolves into larger activity until we slide down the slippery slope and find ourselves in a quagmire. Attention to detail in handling minor incidents will prevent many major incidents while preparing us for the ones we cannot stop (Figure 1.1).

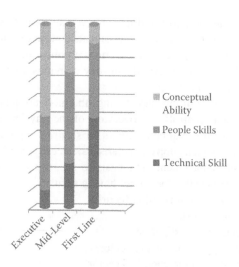

FIGURE 1.1   Chart on management/leader skills.

Why Would I Provide Security?

State/federal/local laws and regulations: For example, 40 U.S. Code § 1315(c) (1) authorizes the Secretary of Homeland Security and the General Services Administration to establish regulations to protect government property. An example of a regulation would be 41CFR105-62.101, which designates security classification categories. Also, Homeland Security Presidential Directive 7 establishes a national policy for federal departments and agencies to identify and prioritize U.S. critical infrastructure and key resources and to protect them from terrorist attacks.

Insurance requirements: Insurance policies may require that certain assets be protected with a minimum level of security. Many homeowners' policies provide a discount if a security system is installed.

Contract requirements: A lease may dictate a level of security in order to meet contractual obligations.

Customer/supplier relationships: A customer may require that a shipper provide additional security when handling packages. Signature required or armored car services are examples of providing additional security.

Union agreements may spell out certain requirements to provide for workplace safety: A labor agreement may require security patrols in employee parking lots or assurances for safety in the workplace against workplace or domestic violence.

EXERCISE

You are assuming a new position of security manager in a relatively new company and you have never before had a security department job. Although some members of management understand the need for a formal security department, others do not. Sure, there have been losses in the company, but the role of security is not going to be without cost, either. It is your first job to meet with the departmental managers and explain to them why their organization needs security. Is security management an investment or just a requirement?

REFERENCES

Homeland Security Presidential Directives (HSPD) 8: National Preparedness. 2011. US Department of Homeland Security, Washington, D.C., March 30.

Northouse, P. 2006. *Leadership: Theory and Practice*. Thousand Oaks, CA: Sage Publications.

2 Security Management and Critical Thinking

The modern security manager should utilize critical thinking in performing the requirements of his/her position. However, the literature of security management and critical thinking methodologies is sparse. Purpura (2002) touches on the utilization of critical thinking when saying that "it is not to criticize, but to be applied discretely to understand the security environment more adequately." Critical thinking is to move the security manager past personal bias, or biases of others, to provide objectivity to situations. Loss prevention practitioners know that security management is rooted in methodical processes that serve as a guide for protection. As such, many loss prevention processes utilize a homogenized type of critical thinking.

The critical nature of security management tools, such as risk assessments, critically and vulnerability studies, return on security program investment, and cost benefit analysis, show just a glimpse in the systematic tools of the security manager. The risk assessment itself looks at the value of organizational assets, accounting for both the criticality and vulnerability of the asset to determine the amount of the security investment. In itself, the comprehensive process of risk analysis utilizes a process of critical thinking.

The intent of this chapter is to move beyond statistical loss prevention tools and to focus on ways the security manager can apply critical thinking skills to improve his/her everyday performance. We will provide an understanding of the critical thinking process as a methodical process that the security manager should cultivate to evaluate the security environment on a daily basis.

For a security manager, there is nothing more practical than a methodical process of thinking. For managers, good thinking pays off while poor thinking causes problems, such as wasting resources and time. Critical thinking allows security managers to envision their duties in a logical process while focusing on making decisions and solving problems.

Critical thinking should become a natural process for security managers. Schon's (1983) idea of critical thinking involves the relations between theoretical and practical knowledge as a part of our daily, spontaneous, and intuitive management processes, which become a reflection process for the security manager to instinctively and critically interpret a situation.

Essential for thinking about how a security manager uses critical reflection is how Schon (1983) describes them as a researcher and practitioner who can theorize and know his/her actions because of reflection. The security manager experiences revelation, perplexity, or uncertainty in situations that he finds to be unknown or unique. The security manager reflects on the situation and on prior understandings that have been implicit on his/her behavior. The security manager instinctively carries out an intuitive, methodical process that serves to generate both a new understanding of the phenomena and a change in the situation.

In the reflective process, the security manager does not separate thinking from doing when coming to a decision, which he/she must later convert into action. The security manager is a practitioner, a researcher, who can engage his/her loss prevention situations by bringing thinking and action together, by interactively and critically engaging the relations of means and ends in situations that are unique, difficult, and uncertain.

Critical thinking and reflection is much more than a cognitive process of analysis. Critical thinking becomes more than a form of technical rationality to a potential loss situation. Instead, critical thinking and reflection becomes an artistic process (Brookfield, 1987, p. 155). During the performance of their duties, security managers may be faced with unexpected and unfamiliar situations to which they need to respond. They need to look at the situation and engage critical thinking so that it becomes a part of the collection of experiences they use to build theories and responses. Security managers need to observe, research, and reflect on processes that impact the workplace. They also need to be able to articulate their reflection-in-action; otherwise they cannot instill the process in others (Schon, 1983, p. 243). All thinking and practices demand a stop-and-think period of time that constitutes reflection. The security manager should hone a methodology, bringing critical thinking, reflection, and action together, which will lead to a superior daily performance.

BACKGROUND OF CRITICAL THINKING

To arrive at a process of critical thinking that could best serve a security manager, examination must be provided for a better understanding of the critical thinking processes. Critical thinking has been expressed and defined in several ways. In the literature, there are several definitions and ways of conceptualizing critical thinking. Chance (1986) defines critical thinking as the ability to analyze facts, generate and organize ideas, defend opinions, make comparisons, draw inferences, evaluate arguments, and solve problems. Tama (1989) defines it as a way of reasoning that demands adequate support for one's beliefs and an unwillingness to be persuaded unless support is forthcoming.

Mayer and Goodchild (1990) say that critical thinking is an active, systematic process of understanding and evaluating arguments. An argument provides an assertion about the properties of some object or the relationship between two or more objects, and evidence to support or refute the assertion. Critical thinkers recognize that there is no one correct way to understand and evaluate influences and that all attempts are not successful.

Although critical thinking receives much attention in the higher education environment, the idea is not new. John Dewey, the proponent of the progressive education movement, is often regarded as the father of the modern critical thinking. In *How We Think* (1910), Dewey calls critical thinking *reflective thought* and defines it as "active, persistent, and careful consideration of a belief or supposed form of knowledge in light of the grounds which support it and the further conclusions for which it tends" (p. 9). Dewey recognizes that what matters are the reasons we have for believing something and the implications of our beliefs (Fisher, 2001).

For Dewey, critical thinking is essentially an active process in which you think things through yourself, raise questions yourself, find relevant information yourself, and solve problems yourself, rather than learning in a largely passive way from someone else. Thus, to develop critical thinking skills, individuals must be active learners in the learning process and they must be required to identify and solve unstructured problems using multiple information sources.

For a security manager, you must be able to deal with the unstructured situations that arise in the work environment. If every problem or issue that managers faced could be dealt with in the same structured response, they would not need to "think," but merely apply their rubric to the situation and then go to the next issue. However, the security environment seldom exists in that scenario. While many facets of security can be addressed through proactive management tools, there is still a huge variation of problems the security manager deals with on a daily basis. Dewey roots critical thinking in the engagement with a problem (1916).

A century later, contemporary critical thinking researchers promote that by using context, elements, and disciplines to gain insight as well as understanding the issues in a deep way helps make better decisions in life (Nosich, 2012). For Nosich, critical thinking allows individuals to more effectively interpret information, keep the important information with us, and actually retain information for future use and reference with more ease over time because the skills provided allow a person to become an active listener rather than a passive recipient of information.

Nosich uses an academic approach describing the technical process of critical thinking to assist improving individual capabilities through critical thinking. His point of view is that, with practicing critical thinking, we will have the following outcomes:

> Whether it is in writing or reading, in analysis or evaluation, in the discipline as well as in your life outside school, critical thinking creates value. It takes effort, especially before you get used to it. But it has clear practical benefits that far exceed the effort. It will produce better answers, better grades, in more courses, in more professions, with ultimately less work, than any alternative. More than that, it gives insight that can make your life richer, by bringing the elements, the standards, and the disciplines into learning to think things through (p. 195).

If all parts of critical thinking are understood and utilized, then the security manager should be checking for accuracy, identifying assumptions, drawing relevant conclusions, and thinking questions out in terms of the fundamental and powerful concepts of the discipline. Critical thinking is reflective. It is different from just thinking. It is metacognitive involving thinking about your thinking.

Critical thinking involves standards used as a measure for thinking. Some examples of standards are accuracy, relevance, and depth. Nosich states that "there are no rules that guarantee our thinking will be correct," so the critical thinker must allow for "self-correcting" by evaluating the reasoning. The three parts of critical thinking emphasized are "asking questions ... that go to the heart of the matter," "trying to answer those questions by reasoning them out," and "believing in the results of our reasoning" (Nosich, 2012, p. 5).

The elements of reasoning (keeping in mind context and alternatives) are, in no particular order because this is a circle, point of view, purpose, question at issue, assumptions, implications and consequences, information, concepts, conclusions, and interpretations (Nosich, 2012, p. 49–60).

- Purpose is having an objective.
- Question at issue can be thought of as the problem being addressed.
- Assumptions are everything you take for granted when you think through something.
- Implications and consequences fall into the area beyond the end of critical thinking.
- Information is the data, evidence, observations used during reasoning. Reliable information from numerous sources should be used.
- Concepts are our understanding of a term or issue and must be made aware.
- Conclusions are our decisions.
- Point of view is the perspective used to address an issue.
- Alternatives are simply other possibilities.
- Context is similar to the setting or background of reasoning.

There are seven standards to thinking critically, which are clearness, accuracy, importance/relevance, sufficiency, depth, breadth, and precision (Nosich, 2012, p. 133–148).

- Clear thinking is when you can state your meaning exactly, when you can elaborate on it and explain it, when you can give good examples and illustrations of it.
- Importance occurs when the information is directly relevant to addressing the problem at hand.
- Sufficiency happens when you have thought about a question until you have reasoned it out thoroughly enough for the purpose at hand, when it is adequate for what is needed, when you have taken account of all necessary factors.
- The goal for depth and breadth is to develop an intuitive feel for when it is important to delve more deeply into an issue and when it is important to look at it more broadly by taking account of other related issues.
- What is precise will always be relative both to the purpose of the reasoning and to the context.

Nosich describes the core process as addressing a question or problem, thinking it through using the elements of reasoning, reasoning out all aspects of the issue through the lens of discipline when appropriate, and monitoring reasoning using the critical thinking standards (p. 169). Evaluation, comparing and contrasting, application, decision making, action, and living mindfully play into the critical-thinking process (p. 172).

"Critical thinking transfers" (p. xxviii). With "time, practice, and commitment," the reader will be able to use the material supporting how to think critically and critical thinking's application to improve not just in school, but in all decision making with improved decisions as a result. However, we must take action or this will not come to fruition.

Whether it is in writing or reading, in analysis or evaluation, in the discipline as well as in your life outside school, critical thinking creates value. It takes effort, especially before you get used to it. However, it has clear practical benefits that far exceed the effort. It will produce better answers, better grades, in more courses, in more professions, with ultimately less work than any alternative. More than that, it gives insight that can make your life richer by bringing the elements, the standards, and the disciplines into learning to think things through (2012, p. 195). So, the material will allow the reader to incorporate critical thinking into daily life experiences with improved decision-making results.

Paul and Elder (2005) argue that critical thinking involves the ability to raise vital questions and problems; to gather and assess relevant information; to use abstract ideas to interpret information effectively; to come to well-reasoned conclusions and solutions, testing them against relevant criteria or standards; and to think open-mindedly within alternative systems of thought, recognizing and assessing their assumptions, implications, and practical consequences.

Paul and Elder (2005) further argue that successful thinkers move more or less sequentially through a standard process of identifying problems, making reasonable assumptions about the nature of the problems, discerning criteria according to which information about the problems can be deemed relevant and well understood, making inferences from the pertinent data, and organizing these inferences into concepts that will help in coming up with a workable solution.

Paul and Elder use these "elements of thought" to create a checklist for students to guide them in their analytical thinking. The benefit of this checklist to instructors and students is that they teach individuals how to analyze a broad range of materials from news articles to chapters in textbooks to government reports to novels and poems.

Students should regularly use the following checklist for reasoning to improve their thinking in any discipline or subject area:

- All reasoning has a purpose.
  - State your purpose clearly.
  - Distinguish your purpose from related purposes.
  - Check periodically to be sure you are still on target.
  - Choose significant and realistic purposes.
- All reasoning is an attempt to settle some question, figure something out, or solve some problem.
  - State the question at issue clearly and precisely.
  - Express the question in several ways to clarify its meaning and scope.
  - Break the question into subquestions.

- Distinguish questions that have definitive answers from those that are a matter of opinion and from those that require consideration of multiple viewpoints.
- All reasoning is based on data, information, and evidence.
  - Restrict your claims to those supported by your data.
  - Search for information that opposes your position and information that supports it.
  - Make sure that all information used is clear, accurate, and relevant to the question at issue.
  - Make sure you have gathered sufficient information.
- All reasoning contains inferences or interpretations by which we draw conclusions and give meaning to data.
  - Infer only what the evidence implies.
  - Check inferences for their consistency with each other.
  - Identify assumptions that lead you to your inferences.
- All reasoning is expressed through, and shaped by, concepts and ideas.
  - Identify key concepts and explain them clearly.
  - Consider alternative concepts or alternative definitions of concepts.
  - Make sure you are using concepts with care and precision.
- All reasoning is based on assumptions (beliefs you take for granted).
  - Clearly identify your assumptions and determine whether they are justifiable.
  - Consider how your assumptions are shaping your point of view.
- All reasoning is done from some point of view.
  - Identify your point of view.
  - Seek other points of view and identify their strengths and weaknesses.
  - Strive to be fair-minded in evaluating all points of view.
- All reasoning leads somewhere or has implications and consequences.
  - Trace the implications and consequences that follow from your reasoning.
  - Search for negative as well as positive implications.

The point of critical thinking and security management is to promote active exploration of ideas through exploration. Critical thinking should be developed in security managers as they move through the process of identifying problems, gathering facts and data about the problem, making reasonable assumptions about the nature of the problems, discerning criteria to analyze the problems, and identifying possible solutions to complex problems and their consequences.

Security managers use tools, such as risk assessments, criticality and vulnerability studies, return on program investment, and cost benefit analysis, to gain an understanding of loss environments. In itself, the comprehensive process of risk analysis utilizes a process of critical thinking. However, the intent of this chapter is to embrace the formal processes as well as to make critical thinking a reflective part of the everyday role of security management.

EXERCISES

1. Summarize five major points made in this chapter.
2. Discuss the essence of this chapter using a metaphor.
3. Explain critical thinking to your neighbor who has a high school education and has not been in the workforce for 15 years. What assumptions did you make (and why) when you were developing each explanation?
4. How might the information you gained from this reading on critical thinking affect you personally and professionally?

REFERENCES

Brookfield, S. 1987. *Developing critical thinkers: Challenging adults to explore alternative ways of thinking and acting.* San Francisco: Jossey-Bass Publishers.

Dewey. J. 1910. *How we think.* Boston: D. C. Heath and Co.

Dewey, J. 1916. *Democracy and education.* New York: Macmillan.

Fisher, A. 2001. *Critical thinking: An introduction.* Cambridge, UK: Cambridge University Press.

Nosich, G. 2012. *Learning to think things through. A guide to critical thinking across the curriculum.* Upper Saddle River, NJ: Prentice Hall, p. 4.

Paul, R. and Elder, L. 2005. *The Thinker's Guide to the Nature and Functions of Critical and Creative Thinking.* Dillon Beach, CA: Foundation for Critical Thinking.

Schön, D. 1983. *The Reflective Practitioner, How Professionals Think in Action.* Basic Books.

Tama, C. 1989. Critical thinking has a place in every classroom. *Journal of Reading*, 33, 64–65.

FURTHER READINGS

Dick, R. 1991. An empirical taxonomy of critical thinking. *Journal of Instructional Psychology* 18: 79–92.

Ennis, R. H. 1962. A concept of critical thinking. *Harvard Educational Review* 22: 81–111.

Erskine, J. A., R. L. Michiel, and L. A. Mauffette-Leenders, 1981. *Teaching with cases.* Waterloo, Ontario: Davis and Hedersen Publishers.

Purpura. P. 2002. *Security and Loss Prevention: An Introduction.* Walthan, MA: Butterworth-Heinemon.

3 Core Competencies

PLANNING

Every organization, small or large, public or private, must have a road map or guide as to where they want to go or believe that they are going. This must start by knowing what has happened in the past, what is occurring in the present, and where you want to go in the future, and this necessitates having a plan. This is also true for security management.

What is a plan? A plan is selecting and relating facts, and making and using assumptions regarding the future in a rational course of action. The planning process, related to rational decision making, is when a course or courses of action are made. Assumptions generally are accepted as true statements or beliefs without proof or demonstration. These assumptions can be considered a bundle of decisions concerning what is known about the past and, thus, will be included in planning for the future by making decisions now. A plan represents expenditures of thought, time, and knowledge for an investment in the future.

The rational selection of a course of action, that is, the making of a rational plan, includes basically the same procedures of those of any rational decision. This means that most, if not all, courses of action must be identified, the consequences of each course (area) must be predicted or known in advance of decision making, and the courses (onward movement in a particular direction) have the preferred results or consequences, which must or should be selected. In order to achieve these results, the planning process should comprise the following five activities:

1. Problem identification or the analysis of the situation
2. Goal setting (desired future state of affairs)
3. Design of courses of action (alternative approaches to goal attainment)
4. Comparative evaluation of consequences (predicting the results of each alternative)
5. Final selection of course of action (decision making)

These five activities create a dynamic process that involves a number of courses of action or methods for generating plans that provide an organization with guides to sustain renewal and change in terms of more effective goal accomplishments. Combining these five activities means that progressive actions are performed by persons in the course of moving the organization from one state or situation to another. Combining these activities creates or makes a process that allows or requires a *flow of interrelated events moving toward some goal, purpose, or end*. "Flow" implies a movement through time in the direction of a consequence. "Interrelated" denotes interaction within the process, and events that are highly relevant one to another. "Events" are changes or happenings that occur at one point or period of time, and may be any of an infinite number of phenomena. "Goals" suggests a human or decision makers' objective, while "purpose" suggests either human objectives or objectives in a nonmaterial sense. "End" implies some conclusion or consequence that may not necessarily be sought or planned by the decision maker. This explanation or definition of process may or may not have consequences intended by humans.

It is certainly time that some goals can be achieved with relatively little planning; however, today where many tasks have become quite complex, more technology is involved, more people want and demand to be informed and participate in what is going to be done, and, with the ever-increasing diversity of personnel, products, and services, planning has become a necessity.

One may question the relevance of planning or a plan drawn from a major concept of classical management writings. One would have a major argument trying to defend not having any plan whatsoever. As for other specific points that it should have and that turn on the nature, importance, and condition of organization for which the plan is drawn up, there could be a possibility of settling them beforehand and to save time and money by utilizing acceptable components of an already known and successful organizational plan.

ORGANIZATION

Organization can be defined as the establishing of a system of effective and structural behavioral or interpersonal relations among individuals with the individuals being differentiated in terms of authority, status, and role with the purpose of achieving some goal or objective. Organizing results in an organizational structure that provides a framework whereby human can favorably unite their efforts.

Classical organization theory is built around four primary concepts that were formalized from the past successful occurrences. They are the division of labor, the scalar and functional processes, structure, and the span of control.

1. The division of labor has been accepted as the cornerstone of the four concepts. From it, the other concepts flow as a natural consequence or effect. Growth within the organization requires the scalar and functional processes to necessitate specialization and departments of function. Organizational structure is always dependent upon the direction that specialization development takes. Finally, span of control problems result from the number of specialized functions under the authority and supervision of a manager.
2. The scalar and functional processes deal with the vertical and horizontal growth of the organization. The scalar process refers to the growth of the chain of command, the delegation of authority and responsibility, unity of command, and the obligation to report. The division of the organization into specialized parts and the regrouping of the parts into compatible units are matters pertaining to the functional process. This process focuses on the horizontal evolution of the line and staff in a formal organization. A strong belief developed from the line-staff relationships that staff areas and positions were not to supervise and have authority over line processes.

3. Structure is the logical relationship of functions in an organization, arranged to accomplish the goals and objectives of the organization efficiently. Structure strongly implies system and pattern. Structure is the mechanism for introducing logical and consistent relationships among the diverse functions that comprise the organization.

4. The span of control concept relates to the number of subordinates a manager can effectively supervise. Are there numerical limitations to the subordinates one person can control? The answer is no. Span should refer to a number of persons, themselves carrying managerial and supervisory responsibilities, for whom the senior leader retains over-embracing responsibility of direction and planning, coordination, motivation, and control. Regardless of the interpretation of span of control, a wide span yields a flat structure; a short span results in a tall structure. In today's world, most leaders accept a wide span.

MOTIVATION AND CRITICAL ANALYSIS

From the four primary principles of classical management (especially, span of control), comes the advent of centralization versus decentralization. Frederick Winslow Taylor (1856–1915), who had an engineering background and is accepted as the father of scientific management, had to deal with one location and one product and how to arrange this organization and its component parts in a more efficient and effect manner to make greater profits. Emphasis was placed on arranging the organization in ways that would allow the workers to assist it. Under this system of management, the organization was primary and the workers were secondary.

In the United States, during the last three decades of the nineteenth century, the Industrial Revolution sprung up and products were not produced in one location and in one facility. Products, such as oil, steel, gas and oil, paint, chemicals, glass, and financial services, were not available only in one location. Automobiles and railroads and other products required numerous component parts. Organizations started exploring decentralization along with their centralized control.

For years, the best example of this type of span of control was the late Alfred P. Sloan at the General Motors Corporation. Sloan's vision was to divide GM into as many parts as could be done consistently, place in charge of each part the most capable executive that could be found, and develop a system of coordination so that each part could strengthen and support the others. In academic circles, this was labeled "decentralized organization with centralized control." The units that exercised major control were finance, policy, and direct decisions that expanded all parts of the organization. Again, this led those in academia to foster a saying that "in any organization, if you don't control your budget and your personnel, you don't control your part of the organization."

Again in the United Sates, during the 1870s, 1890s, and early 1900s, prior to World War I, organizations grew a large number of products that were produced in one or more locations with an extremely large number of workers.

The emphasis by these organizations to develop around classical theories and, in most cases, disregard the human aspect of the organization, led to the neoclassical school of thought. This is a "tall" form of management.

Neoclassical theory embarked on the task of compensating for some of the deficiencies in classical doctrine. The neoclassical approach to management takes the postulates of the classical school, regarding the pillars of the organization as a given. Neoclassical emphasized that these claims were modified by people, acting independently or written in the context of the informed organization: "flat form."

The neoclassical had the support of the behavioral scientists (and later mathematicians) to assist in conducting studies and analyzing existing data. Studies of the repetitive nature of factory work, assembly line (piece) work in time as workers completed and is a requirement and emphasis on how the informal organization affects these workers.

Without transparency and visible communication in an organization, an informal organization likely will develop. Five important aspects of informal organization as it operates within the formal organization include:

1. Informal organizations act as agencies (parts) of social control. They generate a culture based on certain norms of conduct, which, in turn, demands conformity from group members. Diversity of personnel assists in reducing the effects an informal culture can exert over an organization.
2. The form of human interrelationships in the informal organization requires techniques of analysis different from those used to plot the relationship in a formal organization.
3. Informal organizations have status and communication systems peculiar to themselves, not necessarily derived from the formal organization.
4. Survival of informal organizations requires stable, continuing relationships among the people in them. Informal organizations resist change. Change may involve learning a new job and developing new people relationships.
5. The last aspect of analysis that appears to be central to the neoclassical view of the informal organization is the study of the informal leader. A balancing act exists with the informal leader about how to assist and serve the informal group and, at the same time, assist and serve the formal organization. The informal leader and the formal leader must understand that this leadership position was not created by the formal organization.

In today's society, a 40-hour work week, with extended overtime, long vacations and sick time from work, 20- to 30-year retirements for life at any age, and health and sick benefits fully covered are causing problems for both private enterprise and governmental organizations. These areas have and will create problems for future organizational leaders.

In regards to motivating employees in any organization, two factors are important today and will continue to be important in the future:

1. Communication and transparency are important to all employees. The more they know about what is going on in their organizations, the more they relate to the organization.

2. The involvement of employees at all levels and parts of the organization in organizational decision making will assist in the motivation aspects of employees. Authentic employee involvement must be the organizational norm. Employees expect that every suggestion they make for improving the organization will be considered. They want involvement and suggestions to be discussed with them and explanations given as to whether or not they will be implemented. Employees are not fools, and they do not want hypocrisy from management.

MOTIVATION

Learning theories, as they relate to individual and group behaviors, have evolved and developed since World War II. Drastic changes following the war opened up many opportunities for social science researchers to publish research work and studies, such as the Hawthorne effect study (Hawthorne Works, Chicago, 1924–1932), from the years prior to the war, and explore these ideas and even develop new ideas for the now fast-paced society.

Two major researchers, A. H. Maslow (1954) and Fredrick Herzberg, during the same time period, developed Models of Motivation:

Model of Maslow's hierarchy of needs
    Self-actualization and fulfillment
    Esteem and status
    Belonging and social needs
    Safety and security
    Physiological needs
Model of Herzberg's motivation—maintenance (Herzberg et al., 1959)
    Work itself, achievement, possibility of growth responsibility
    Achievement, recognition, status
    Relations with supervisors, peer relations, relations with subordinates, quality of supervision
    Organization policy and administration, job security, working conditions
    Pay

These two models are cognition (cognitive) theories of motivation that relate to the mental process by which knowledge is acquired. The models research how a worker obtains this knowledge through perception, reasoning, and intuition. A major difficulty with cognitive models of motivation is that they are not subject to precise scientific measurement and observation. One must argue that these two models are strong starting points if any type of reinforcement to modify behavior by its consequences are in place. One also could argue that most people today have satisfied their lesser needs and are looking toward satisfying higher-level needs at work. Satisfying higher-level needs can be done in most organizations by decreasing hierarchical control and overspecialization of roles. Many studies have discovered that decreasing hierarchical control leads to a greater self-value, and, hence, a greater degree of expectancy of success and value attainment.

CRITICAL ANALYSIS

Defining or stating a problem must be the first order of business when analyzing a critical problem or potential problem before making a decision. The core or the central part of the problem must be stated before an analysis starts and a decision is made. After stating the problem, decisions must be made as to what kind of data must be collected and, then, set priorities on how a decision will be made.

In social science, we use a number systems for essentially three purposes: (1) to classify things, (2) to order things, and (3) to quantify things. Because data are measures of such variables, we refer to them as (1) nominal data, (2) ordinal data, (3) interval data, and (4) ratio data.

Nominal data is whenever we assign numbers to a set of categories without reference to direction or magnitude of difference among the alternatives. Nominal pertains to, or consists of, a name or names.

An ordinal scale includes the essential property of a nominal scale, plus two more:

1. The categories are mutually exclusive.
2. They are ordered according to the amount of the attribute they represent.

An interval scale is characterized by three basic properties, including the two listed above.

1. The categories are mutually exclusive.
2. They are ordered according to the amount of the attribute they represent.
3. Equal differences in the attribute are represented by equal differences in the number assigned to the categories.
4. Numbers assigned to the categories are proportional to the amounts of the attribute represented by them.

A ratio scale has all of the attributes of an interval scale with the added difference:

1. Both interval and ratio levels are quantitative measurement. The difference between an interval and a ratio scale is that the interval scale has an *arbitrary* zero point while the ratio scale has an *absolute* zero point.

The use of the quantitative and/or qualitative measures is also an important consideration in planning for the data collection phase of critical analysis. Quantitative measures are typically numerical, and qualitative measures are textual. Quantitative variables are aggregate indicators of the magnitude of concepts, whereas qualitative variables show the character or content of concepts. It is important to keep in mind that neither one variable is better or worse than the other. The use of either type of data should be consistent with the research questions one is asking and not because one has a preference for statistical or qualitative analysis.

Three kinds of "averages" are defined from data that have been obtained for research questions. The "mean" is the point around which the values in the distribution balance, and is the mathematical or arithmetic average. In order to calculate a mean, internal level data must exist. The median provides information about the value of the middle position in the distribution. It is the point in the distribution of values at which 50% of the scores fall below and 50% fall above. In order to calculate a median, you must have at least an ordinally measured variable. Mode represents the most frequent value in distribution. The mode is the simplest measure of averages and, therefore, is not viewed as an overly precise or informative measure of average. In addition, the mode is the only measure that is appropriate for the nominal learned of measurement.

Security organizations, small or large, public or private, must have a road map or guide as to where they want to go or believe that they are going. This must start by knowing what has happened in the past, what is occurring in the present, and where one wants to go in the future. This necessitates having a plan. This chapter has shown, through classical studies, how these theoretical principles are true for security management as well.

EXERCISES

Maslow's hierarchy of Needs is a core competency that a security manager should understand. It can be used by the manager to understand how employees interact in life and also in the work environment. Explain how Maslow's hierarchy can be used to explain internal theft by employees. How would this explanation be different from using Maslow's theory to explain theft from outsiders?

REFERENCES

Maslow, A. H. 1954. *Motivation and personality*. New York: Harper and Row.
Herzberg, F., B. Mausner, and B. B. Snyderman. 1959. *The motivation to work*. New York: John Wiley & Sons.

4 Developing a Security Management System

Security management is a broad field of management related to asset loss prevention, physical security, occupational safety, and intangible asset protection functions. It entails the identification of an organization's employees, physical assets, intangible assets, and the development, documentation, and implementation of physical measures, policies, procedures, and guidelines.

Security management is a methodical process that is used to develop mechanisms to protect organizational assets. Security management tools, such as risk assessments, criticality and vulnerability studies, and cost benefit analysis show just a glimpse of the methodical tools of the security manager. The risk assessment itself looks at the value of organizational assets, accounting for both the criticality and vulnerability of the asset to determine the amount of the security investment. The comprehensive process of risk analysis, as well as other security management tools, utilizes a process of critical thinking, while providing a basis for a comprehensive security management program.

UNDERSTANDING THE ORGANIZATION TO ESTABLISH SECURITY PROGRAMS

The role of security management must begin by asking some fundamental questions in determining where to begin in the process of protecting organizational assets. The findings, provided by a critical thinking framework, are based on the analysis of several security-related factors, which serves as the basis for the implementation of security measures to develop an optimal level of protection. The desired level of protection is the degree of security provided by a particular countermeasure or set of countermeasures to protect the asset.

IDENTIFY ORGANIZATIONAL ASSETS AND GOALS

The first step of the security management planning process is to identify organizational assets and specific overall goals for the organization. This part of the planning process should include a detailed overview of each organizational asset and its relationship to organizational goals, including the reason for its selection and the anticipated outcomes of goal-related projects. Where possible, objectives should be described in quantitative or qualitative terms. These quantitative and qualitative goals should be measurable. Being able to measure needs, as well as outcomes, is fundamental to security management as well as to the entire organization.

The application of security programs shall follow the critical thinking process regardless of the type of asset involved. The asset has a direct impact on the application of this process and how countermeasure recommendations resulting from the process are to be implemented. Application of this security standard ensures a comprehensive approach to meeting organizational security needs in the threat environment, and that the scope of security is commensurate with the risk posed to an asset, relative to cost.

Each goal should have financial and human resources considerations associated with its achievement. For example, we have an organization that manufactures widgets and if the organization has a goal to increase production and sales by 20%, how does that impact the role of security? What aspects of organization assets, risks, and vulnerabilities change when the overall goals of the organization change? Maybe assets, risks, and vulnerabilities don't change. However, we do not know that until we adequately understand the organizational environment.

Therefore, a primary goal in security management is to identify and evaluate organizational assets. For simplicity, we can categorize assets as people, physical, and intangible.

People

The people category can range from the line workers providing minimal strategic importance to vital individuals holding key roles, whose incapacity or absence will affect the business.

Physical Assets

This includes all physical organizational assets, including real estate, buildings, facilities, equipment, materials, moneys, supplies, inventories, and all the physical resources that allow the organization to operate.

Intangible Assets

Intangible assets include information, plans, and organizational strategies. It can be marketing and sales plans, detailed financial data, trade secrets, personnel information, sensitive office correspondence, and minutes of meetings. It also can include things such as a positive public perception of an organization. Often these organizational assets are overlooked at first glance. However, the security manager, as well as organizational leadership, should work together to illuminate and assess the significance of these assets.

RELATIONSHIP BETWEEN ORGANIZATIONAL ASSETS

The security manager must understand the relationship of assets in organizational structures and functions. This includes the physical and logical relationships that assets have with each other in the organizational environment. It also can relate to the impact one asset can have on the other in a loss event. An essential tool for a better understanding, the relationship between organizational assets is through the configuration of an asset hierarchy. A very simplistic asset hierarchy diagram for our widget-making machine may look like Figure 4.1.

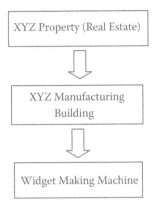

FIGURE 4.1 Asset hierarchy.

From this illustration, we must assume that a loss event impacting XYZ property (real estate) may impact the building and, subsequently, the widget-making machine. For example, if the property floods, that may result in the building flooding as well, which, in turn, may result in loss to the widget-making machine. In another example, a loss to the XYZ Manufacturing Building asset, such as a fire, also may have an impact on the widget-making machine. Or, in a final situation, an employee could misuse the widget-making machine and make it inoperable, which results in a loss to the organization. Therefore, from these scenarios, we can see that there can be a relationship where one asset loss is attributed and can be related to the hierarchical relationship that asset shares with other organizational assets (Figure 4.2).

The point of the asset hierarchy is not as much about making a rigid model of what or where the assets resides in the organizational structure, but, rather, to enlist the asset hierarchy as a critical thinking tool to better understand the physical and logical relationships that assets have with each other in the organizational environment and the impact that one asset can have on the other in a loss event.

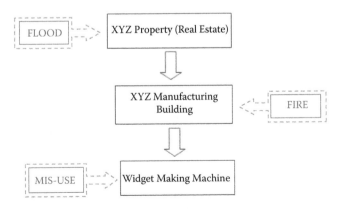

FIGURE 4.2 Asset hierarchy with risk/threats.

IDENTIFY THE CRITICALITY OF ASSETS IN MEETING ORGANIZATIONAL GOALS

One must determine the asset criticality for the organization. This evaluation is where the individual assets are linked to the relationship of the organization's strategic plan. In the organizational environment, Security must understand the value (or criticality) of an asset in attaining organizational goals. If the organizational goal is to increase widget production by 20% in the following year, what is the role of the assets in goal attainment?

A simplistic way to look at the criticality of organizational assets is to place a criticality value on that asset to provide a numeric valuation system, which expresses the value of the asset to the organization. For example, we could lump assets into three general categories: critical, not critical, and no criticality value. From these three general views, the security manager also can include redundancy criteria in the asset criticality measurement. From this scenario, the security manager can rate organizational assets on a scale of 1 to 5 for criticality, with 5 being the most critical (with no redundancy) and 1 having no criticality in meeting organizational goals:

   5: Critical asset, no redundancy
   4: Critical asset with redundancy
   3: Not critical asset, no redundancy
   2: Not critical asset with redundancy
   1: The asset has no criticality to organizational goals

Therefore, an asset that is critical to the operation and has no onsite replacement would be given a criticality of 5, one that is critical to the operation and has a replacement onsite would be given a criticality of 4, and so on.

The premise in understanding organizational assets and their relationship in meeting organizational goals is crucial because failure to fully understand organizational assets will result in the over or under protection of those assets. The resulting over protection of assets will cost the security manager resources that may be better used elsewhere, while the under protection of assets can contribute to greater exposure to risk and subsequent loss.

EXISTING LEVELS OF ASSET PROTECTION

Once all organizational assets are evaluated, the next step is to determine their existing level of security protection. Depending on the organization's requirements, assets may be classified into two or more levels of security need. The security manager should be cautious of having too many existing levels of security protection; this tends to dilute their importance as well as over complicating security management. Having too many security levels proves expensive in terms of employee training, security resources, and practices, in that the costs are often greater than the potential losses. Keep it simple in providing adequate protection while also minimizing complexity in application.

The security manager can utilize the level of protection concept to gain an understanding of security needs by forming needs into three general categories:

1. Existing level of protection: The degree of security provided by countermeasures to an asset currently being utilized at an organization.
2. Necessary level of protection: The degree of security provided by a set of countermeasures identified, which must be implemented and justified by a risk assessment group to provide a necessary level of protection.
3. Optimal level of protection: The degree of security needed to completely mitigate organizational risks.

The existing level of protection may be determined through site surveys, interviews, reviews of policies and procedures, testing, etc., to determine what countermeasures are currently in place and how effective they are. Current conditions may then be matched up against the existing level of protection and then compared to the necessary level of protection to determine if it adequately addresses the threat(s) or if vulnerabilities need to be addressed.

In determining existing levels of protection, the security manager should define, in detail, the following key areas of existing security management:

- Existing asset classification practices: Guidelines for specifying security.
- Previous risk assessments and organizational understanding of asset's risk.
- Assignment of organizational asset ownership: Is there an assignment of roles for handling organizational assets?
- Existing asset responsibilities: The tasks and procedures to be followed by the entities handling the asset.
- Existing policies regarding mishandling of organizational assets: Includes how security violations are reported and dealt with.
- Existing security awareness practices: Education programs and classification of assets.
- Existing security audit procedures: Unannounced checks of security measures put in place to find out whether they are functioning.

If the existing level of protection equates to the necessary level of protection, current countermeasures should be maintained and evaluated on a regular basis. However, organizational conditions should be monitored for changes that may impact the effectiveness of countermeasures or the needed level of protection. Therefore, if the existing level of protection does not sufficiently address the risks, shortfalls must be identified and countermeasures to address those vulnerabilities must be considered for implementation.

ANALYZING RISKS TO ORGANIZATIONAL ASSETS

An effective security management system demonstrates a careful evaluation of how much security is needed to protect organizational assets. Security managers must realize that too little security means that organizational assets can be easily compromised, while too much security can make assets hard to use or so degraded that performance is negatively affected. Security must be inversely proportioned to an asset's utility. It is a given that there is always going to be risk associated with assets and activities. The only way to completely eliminate the risk, in many cases, would be to make that asset inoperable. Therefore, the role of security management is to find the optimal relationship between organizational processes, assets, and functionality. While, the risk assessment process will be more deeply analyzed in Chapter 9, it still maintains merit to look at the overall role in building a security management program.

If the organizational asset is viewed through the inventory of potential loss events, the security manager must recognize that findings are not necessarily all-inclusive. For each undesirable event where the assessed risk is either less than or exceeds the baseline level of protection, the security manager must identify the countermeasures that will provide a level of protection equivalent to the level of risk. For a lower-level risk, minimum countermeasures are typically less stringent, but also may be less effective in mitigating higher risks, while, at the other extreme, very high countermeasures are typically more stringent and, generally, more effective.

A minimum level of risk should be alleviated with minimum countermeasures, a low level of risk should be lessened with low-level countermeasures, and so on. Adjusting from the baseline level of protection by determining which countermeasures are applicable to the assessed risks and identifying changes from the baseline level of protection, the necessary level of protection can be determined. Once the level of protection necessary to meet the risk is identified, an evaluation of current conditions must be made to identify the existing countermeasures.

Security managers should understand that some degree of risk is assumed with all organizational activities and assets. Having no risk is virtually impossible even with unlimited resources. Therefore, identify all risks to organizational assets, then determine risks to accept and others to mitigate via security measures. The security manager must work with others throughout the organization to understand the effect to the business if an asset is lost or compromised. By doing this, you get a good idea of how resources should be assigned to protecting the asset.

For security managers, risk management is a fundamental component of their position. Risk management is the comprehensive approach to allocating resources for the protection of assets to achieve an acceptable relationship between risk and protection. Risk management decisions are based on the application of risk assessment, risk mitigation, and risk acceptance. The risk assessment itself looks at the value of organizational assets, accounting for both the criticality and vulnerability of the asset to determine the amount of the security investment. The comprehensive process of risk analysis utilizes critical thinking. The risk assessments will be the basis for management of risk through the application of strategies and countermeasures to reduce the threat of, vulnerability to, and/or consequences from an undesirable event. Risk acceptance is an explicit or implicit assumption that some risk is not feasibly possible to mitigate.

The desired levels of protection of assets should be critically determined via a risk-based analytical process or risk assessment. The process will focus on risk as a measurement of potential harm or loss from an undesirable event. Understanding risk means understanding threats, vulnerabilities, and consequences. The level of risk is the combined measure of threats, vulnerabilities, and consequences posed to assets from specified loss events.

If the existing level of protection is insufficient, a determination must be made as to whether the necessary level of protection can be achieved; specifically, if the countermeasure can be physically implemented, and whether the investment is cost effective. Cost effectiveness is based on the investment in the countermeasure versus the value of the asset. In some cases, investment in an expensive counter-measure may not be advisable because the life cycle of the asset is almost expired. Additionally, consideration should be given to whether other countermeasures may take priority for funding. Note that cost-effective is a different determination than cost-prohibitive. A countermeasure is cost-prohibitive if its cost exceeds available funding. Funding may exist for a countermeasure, but it may not be a sound financial decision to expend that money for little gain, making it not cost-effective.

There are a variety of mathematical models available to calculate risk and to illustrate the impact of increasing protective measures on organizational assets. For the purposes of this example, the assumption is made, at this step of the process, that there are no countermeasures in place and complete vulnerability exists. This approach is necessary to ensure that all security criteria will be considered as the process is completed, and to define the relationship between the level of risk and the level of protection. The level of risk must be mitigated by a commensurate level of protection. Often a high level of risk must be mitigated by implementing a high level of protection.

Risk assessments utilize a form of critical thinking. Glaser's (1941) premise that critical thinking is (1) an attitude of being disposed to consider in a thoughtful way the problems and subjects that come within the range of one's experiences, (2) knowledge of the methods of logical inquiry and reasoning, and (3) some skill in applying those methods is very applicable in conducting risk assessments.

A very simplistic example of a risk assessment, utilizing critical thinking, could be evaluated by taking the asset value multiplied by the criticality of the asset multiplied by the vulnerability of the asset to determine an actual risk value. The equation is exemplified as $AV \times C \times V = RV$. To make this equation operational, the security manager can first take an asset and determine its value. Then multiple it by its criticality value, which is represented as a percentage from .0 to 1.0. Next, multiply by vulnerability, which is also a percentage expressing the likelihood that a risk event will materialize into a loss.

The following is an example of a critical thinking risk assessment: $AV \times C \times V = RV$:

Asset value (AV): AV considers an asset's value as the original cost, adjusted upward for improvements since the purchase of the asset and downward for loss in value related to the aging of the asset (Damodaran, 2012).

Asset Criticality (C): C in the equation represents the value or degree of importance of an asset in regards to achieving organizational goals. For this value, the security manager can apply asset criticality as a percentage.

5	Critical asset, no redundancy	100% or 1.0
4	Critical asset with redundancy	75% or 0.75
3	Not critical asset, no redundancy	50% or 0.5
2	Not critical asset with redundancy	25% or 0.25
1	The asset has no criticality to the organization	0% or 0.0

Asset Vulnerability (V): V is the likelihood that a risk/threat event will occur to an organizational asset. Vulnerability is based on past loss data, which may be self-collected, or historical statistics from the organization's archived data or data from like organizations. The asset vulnerability would be defined as a percentage of likelihood that a specific loss event will occur in the budget period.

Risk Value (RV): RV is the estimated cost of a risk to an organizational asset determined by multiplying the asset value by the assets criticality to the organization, multiplied by the likelihood (vulnerability) of the occurrence of the risk event, resulting in the valuation of the risk in monetary values.

The following is an example applying the Risk Value Model (RVM) to a scenario as an aid to demonstrate risk value to an organizational asset. To simplify it, the asset will be our widget-making machine, which costs our organization $10,000. The widget-making machine is critical to our organization, but we do have redundancy in that we always keep an extra machine on hand. Based on past loss data, the security manager knows that twice in the past 10 years the widget-making machine was stolen and not recovered.

AV = Our asset value for the widget-making machine considers asset value as the original cost, adjusted downward for loss in value, because it is two years old and has a life expectancy of 10 years. Therefore, AV = 10,000 × .80 (remaining life expectancy or $8,000.

C = The widget-making machine is deemed a critical asset, but we do have redundancy. Therefore, 0.75 represents the value or degree of importance of the widget-making machine in achieving organizational goals.

V = The likelihood that a theft will occur to the widget-making machine is 20% annually. This assumption is based on past loss data, in that the security manager knows two of the machines were lost to theft in the past 10 years.

RV = From this exercise, the annual risk value of theft to the widget-making machine: $8,000 (AV) × .75(C) × .20(V) = $1,200.

So does that mean we can spend $1,200 annually to prevent the machine from being stolen? No, not necessarily. What it does mean is that we have utilized a critical thinking framework to give us a better understanding of our loss environment to an organizational asset. Based on this exercise, we can fiscally justify a system to protect the machine against theft. The security system to protect the machine may include bolting it to the floor, using measures permanently identifying it a property of XYZ Company, assigning unit or staff ownership/responsibility of the machine, or directing policies and rules facilitating an increase in security to the widget-making machine.

The idea in the Risk Value Model is to provide the security manager with a critical thinking tool to protect organizational assets. In practice, various risk assessment methodologies will provide varying outputs, from numbers and percentages to qualitative ratings. The security manager must determine what outputs from their respective methodologies correlate with the desired level of protection.

In an organization where multiple risk assessments may be conducted, the security manager will need to evaluate the comprehensive findings and determine what countermeasure recommendations to implement, or if a single risk assessment will be accepted for application. In gauging the value of a risk assessment, it should meet the following criteria at a minimum:

- The methodology must be credible, and assess the threat, consequences, and asset vulnerability to a specific loss event.
- The methodology must be reproducible, and produce similar or identical results when applied by various security professionals.
- The methodology must be defensible, and provide sufficient justification for deviation from the baseline.

For the security manager, the risk assessment is a fundamental component of determining protection efforts. The risk assessment can be a critical thinking tool utilized in allocating resources for the protection of assets. For the security manager, risk management decisions are based on the application of risk assessment.

COST VERSUS BENEFIT OF SECURITY

Cost considerations can be a primary factor in a decision to implement a countermeasure. A cost-benefit analysis is a critical thinking tool used by security managers in justifying security measures or programs. Cost-benefit analysis is a cost analysis methodology used to justify security expenditure; however, all costs, including life-cycle costs, should be considered in whatever methodology is used. In addition to direct project costs, those expenditures associated with indirect impacts (e.g., business interruption, loss in productivity, or loss in credibility) should be considered. Any decision to not secure assets or to defer implementation of security measures due to cost (or other factors) must be documented, including the acceptance or transfer of risk.

The cost-benefit analysis is a comparative assessment of the benefits from your security measure and the costs to perform it, in relation to the financial impact resulting from potential loss to the asset. In a cost-benefit analysis, everything gets a dollar value. The security manager should consider costs for all phases of the security project. Costs may be one-time capital or recurring costs, such as personnel time, supplies, materials, or maintenance. The determining criteria in a cost-benefit analysis are that the security benefit exceeds the cost. How much the benefit exceeds the cost in determining security measure implementation is going to be based on the security manager in consultation with organizational leadership. The premise is that the cost-benefit analysis is another tool used by security management to critically deal with risk to assets.

IMPLEMENT THE SECURITY PROGRAMS

When the security manager completes the critical analyses via asset criticality and vulnerability studies, risk assessments, and cost benefit analysis, then he/she may implement the security measures that have been determined to best fit the asset risk. Implementation of new security programs are best accomplished in stages to make it easier for the organization to adapt to changes in the working environment. The security manager and organizational management should understand that there may be user resistance to security functions. It is recommended that staged implementation be performed starting with the most critical or vulnerable assets.

ASSIGNMENTS AND TIMELINES

As the organization implements security measures, it must establish timelines for completing the associated tasks. This portion of the security implementation process should consider the capability of staff members and the time necessary to realistically complete projects. The amount of preparation required to implement security measures may limit its immediate achievability. If the security measure has no capital cost, such as policy and procedural changes, or can be incorporated into a new project, the countermeasure often can be implemented immediately. When countermeasures require advance budgeting or coordination with outside vendors, implementation may be delayed.

MONITOR FOR COMPLIANCE

Effective security management depends on adequate compliance monitoring. Most often violations of security practices, whether intentional or unintentional, become more frequent and serious if not detected and acted upon. Compliance monitoring has two primary activities: detecting security violations and responding to them.

The security manager should document the response to violations, and follow up immediately after noncompliance is detected. The organization should have a designated response group to deal with security violations. Members of the response group should have access to organizational leadership so that severe situations can be dealt with immediately.

A critical part of noncompliance should be the generation of reports for organizational leadership that discusses security violations. An additional objective of monitoring security measures for noncompliance is to identify potential security violations before they dilute the effectiveness of the program or before they cause serious damage.

REEVALUATE ASSETS AND RISKS

Security management is a discipline that should be dynamic. As changes in the organization or assets occur, a reassessment of the security measures should occur as well. Organizational leadership should keep security management abreast of larger changes in the organization so that security operations and measures are prepared to meet these challenges.

Even the best-laid plans can sometimes be changed by unanticipated events. A security management plan should include contingencies if certain aspects of the master plan prove to be unattainable. Alternative courses of action can be incorporated into each segment of the planning process, or for the plan in its entirety. The security manger must continually identify and analyze threats and vulnerabilities to assets, recommend and implement, and take appropriate countermeasures.

CONCLUSION

Developing a Security Management Program requires a broad field of knowledge in asset loss prevention, physical security, occupational safety, and intangible asset protection functions. It requires a comprehensive knowledge of organizational assets and the development and implementation of physical measures, policies, procedures, and guidelines to protect those assets

Security management requires critical thinking skills in developing mechanisms to protect organizational assets. The process of security management utilizes processes of critical thinking, which provides a basis for a comprehensive security management program. The dynamic nature of organizations and environments require that the security response also be dynamic.

EXERCISES

Is security an expense or an investment? Compare and contrast the expense investment relationship and explain this in detail.

REFERENCES

Glaser, E. M. 1941. *An experiment in the development of critical thinking.* New York: Bureau of Publications, Teachers College, Columbia University
Damodaran, A. 2012. *Investment valuation: Tools and techniques for determining the value of any asset,* 3rd ed. Hoboken, NJ: John Wiley & Sons.

5 Policies and Procedures

Organizational vision and mission are further defined by policies and procedures. The security unit must develop policies and procedures to promote the vision and mission while protecting organizational assets. A policy tells us *what* to do while a procedure tells us *how* to do it.

For this chapter, commonly used definitions include:

Policy: The organization's guiding or governing principle

Procedure: Tells people how to do something; may comprise several work instructions

Protocol: Another name for procedure; associated with specific disciplines

Process: A series of interrelated activities that result in an outcome; comprises several procedures

SOP: Standard operating procedure

Work Instruction: Specific steps within a procedure; assumes one person or job completes the task from start to finish; can be used for training; usually unit contained or job specific

A policy is the organization's guiding or governing principle. It is a general guideline that sets forth parameters for decision making along with the authority for implementation. Policy development is a planned process requiring due diligence in thought because a policy becomes a strong principle of the organization. Planners must look to the desired outcomes while considering the needs of end users. Policy justification should reflect problem solving while demonstrating persuasive reasoning, clarity, and coherence. How will the policy be enforced? Is there an incentive (including negative incentives)? Is there support for the policy from the workforce? Is there some overriding need for the policy? The policy should be consistent with the organization's mission, culture, strategy, and vision. It should not overlap or contradict with other policies and procedures.

Clarity is important in writing a policy. The average reader should not have any questions about the objective of the policy. When writing, do sufficient research to assure you are compliant with other policy and legal requirements. All stakeholders should have an opportunity to have input and give feedback about the policy. In a large organization, worker committees or teams may be organized to assure feedback is received from all levels. When written, policies must be published in a form that is easily accessible to those affected by the policy. In general, a policy manual should be easy to use with a format that allows for ease of updating.

A policy or procedure should be written (or revised) when there is confusion about what appropriate action to take in a given situation, for fairness, consistency, and accountability. Sometimes, legislative or regulatory changes require a change in policy or procedure. After some time, a policy or procedure may have so many exceptions, exemptions, or waivers that the policy (or procedure) is ineffective. This requires a review and either a rewrite or enforcement of the present policy (procedure).

Not all actions require or fall within a written policy. While policy sets out the order of business, anomalies occur and are handled on a case-by-case basis. Policy makes for efficient work. We may be tempted to write a policy for every possible event and become inefficient by spending valuable resources writing nothing but policy, and you end up with a policy manual so large it won't be read. The manual becomes an organ to catch people doing wrong rather than direct people on doing right.

The policy should begin with a policy statement, which explains what you are doing. Is the policy a standard or a guideline? Standards are specific requirements that must be met while guidelines identify best practices. The statement tells when the policy should be applied and lists major conditions or restrictions. Next, document the reason for the policy and supply a description of the conflict or problem the policy is designed to resolve. Have a list of those who should know this policy and should follow the policy and accompanying procedures in order to do their job. Also, document the resources and contact information at the end of the policy.

Procedures tell us how to do something. They are the steps for getting things done. Inside of a procedure are work instructions; specific steps within the procedure. Work instructions usually involve one person accomplishing a task. The aim for a procedure is to use one action per step and to assign the action to a person who is responsible for the action. Many of the general considerations of procedures occur when writing policies. In lieu of repeating material, just keep in mind the ideas presented above.

A procedure begins with a purpose statement. What are you trying to accomplish? Then list the actions and sequence, who does each step to include handing off to another person, where and when this must be done, and any standards for completing the work.

When a new policy or procedure is adopted, the work staff must be educated as to its existence. In the case of a procedure, this may require training. Whether through a training session or a briefing, notice should be presented in a formal format, with documentation of all persons receiving the new material. New employees should be notified of policies and procedures in new employee orientations.

When the policy, or procedure, is in effect, enforcement is usually through disciplinary action. Discipline must be consistent or the policy (procedure) will lose its power. Depending on the severity of a policy violation, progressive discipline provides a level of punishment commensurate with the violation. For example, a first offense may result in a one-day suspension; a second offense, a one-week suspension; and a third violation, termination. There are cases where two employees might get a different punishment for the same conduct. Discipline may be more severe for a seasoned veteran than a new employee. Another consideration may be the seniority of employees. Some policy violations are so severe as to warrant immediate termination. These are usually reserved for those violations that could result in death or serious injury to others, or on matters of integrity. Embezzling money from the organization may not create a risk of injury, but the nature of the violation would warrant dismissal.

When a policy (procedure) has not been enforced, or has been selectively enforced, reviving the policy (procedure) follows basically the same notification process as a new or rewritten policy (procedure). When reviving a policy (procedure), personnel are informed that the policy (procedure) had previously not been enforced, but that it is still applicable and necessary to the function of the organization, and that, forthwith, the policy (procedure) will be enforced. Consideration should be given to a graduated disciplinary process in reimplementing the policy (procedure).

There are two types of security policies: policies affecting the workforce and policies specifically related to the security guard force. Policies affecting the workforce are as varied as the imagination allows. Some of the most common are policies affecting access control, company equipment, visitors to the workplace, and information security. Policies affecting the general workforce are often seen as restrictive and a hassle. It is important for the policy to state the reason for the policy and that the policy is applied throughout the organization. If the chief executive officer (CEO) doesn't have to comply, employees begin to question why they have to comply. Executives should set the example by following all policies and directing others to do so.

SECURITY POLICIES THAT AFFECT THE GENERAL WORKFORCE

Access Control

A policy on "site access control" (not IT access control) should state reasons for buildings and doors to be open or locked, and the time doors will be open or locked.

> The administration building doors will be unlocked Monday–Friday from 7 a.m. to 7 p.m. Access at other times is by key or contact security.

The policy should dictate who is entitled to a key and a log maintained of all keys issued. The policy should require Human Resources to withhold final pay until keys are returned, e.g.:

> Keys are issued to all full time employees for their office. Keys shall not be duplicated and are to remain in the possession of the employee at all times. Keys shall be returned upon resignation or termination.
> Supervisors are authorized a master key for all offices in their area. Master keys shall not be loaned out and shall remain in the possession of the supervisor at all times.

Electronic access control systems should dictate times and days personnel are allowed to enter the premises. Access control cards and fobs shall be returned on resignation or termination. Employees should be aware that their card may be deactivated for security reasons, e.g.:

Employees are issued access control badges that shall be displayed on the person at all times while on the property. Badges should be visible above the waist by hanging on a lanyard around the neck or clipped to a shirt, blouse, or jacket. Employees are not to loan the badge to another person ... Misuse can result in having your badge deactivated.

VISITORS

Visitor control policies should state the process for permitting visitors access to the facility and who has permission to sponsor a visitor, e.g.:

Any employee may sponsor a visitor to the facility. Visitors must be met at the front desk by the employee who will sign for the visitor badge and escort the visitor at all times while he/she is in the facility. The employee is responsible for signing out the visitor when he/she leaves the facility.

Visitor policies also may be applied to direct areas where visitors are not allowed access, e.g."

No visitors are allowed in the research lab at any time without the written approval of the lab director.

A visitor policy may contain a procedure employees will follow if they find an unidentified visitor, e.g.:

Any employee who observes an unescorted visitor or unknown person not displaying an employee badge shall immediately notify security and provide a description of the person. The employee is not to approach the unknown person. The employee may be asked to watch the person until security arrives and to notify security if the person moves to another location.

COMPANY EQUIPMENT

A policy for equipment control should state the authorizations needed to remove property from the premises, e.g.:

All company equipment is marked with an identification tag and must be checked out before leaving the facility. The employee must have the approval of a supervisor to remove property from the facility.

INFORMATION SECURITY POLICY

Information security, whether hardcopy or digital matter, should be secured by users and steps taken to guard against theft or compromise, e.g.:

All documents shall be secured at the end of the workday by placing all files in the workroom. The workroom shall be closed and locked at the normal close of business.

Documents shall not be left exposed on any desk or workspace. If you need to leave your work area briefly, use a coversheet to protect the documents from view.

Computer screens should be positioned to protect against view from passersby. When leaving your workstation, computers should be placed on standby with a password to reopen.

POLICIES AFFECTING THE SECURITY FORCE

JURISDICTION AND AUTHORITY

The Security policy should include a policy identifying the jurisdiction and authority of the guard force, e.g.:

XYZ, Inc. guards have no authority to arrest. State law [*cite statute*] authorizes guards on premise to detain suspicious persons for law enforcement authorities. Guards have no authority or jurisdiction to act when off company property.

CODE OF ETHICS

The security force should adopt or develop a code of ethics. All personnel in security operations shall receive training on the code of ethics and agree to abide by the code, e.g.:

All security personnel will receive [_] hours of ethics training. At the completion of the training, each employee will sign an agreement to abide by the code of ethics.

USE OF FORCE

The decision to allow or not allow force by guards should be stated with details on the use of force allowed or authorized by law, e.g.:

Guards are not allowed to use force except to prevent injury to another person. The amount of force permissible is dictated by state law [*cite statute*]. At no time is deadly force authorized to protect property.

When lethal or nonlethal weapons are issued or allowed, the policy should dictate the use of the weapon, e.g.:

Firearms are issued for the personal protection of the guard and personnel onsite. Deadly force is authorized by most state law in the following instances: Only under conditions of extreme necessity

As a last resort

When all lesser means have failed or cannot reasonably be employed

Pepper spray is issued to guards for personal protection. Pepper spray is allowed when [...]

If a guard discharges his or her pepper at an assailant, the guard must immediately inform the assailant that he/she has been exposed to [...]

The use of force policy should require a report on any use of force, e.g.:

When guards must resort to force in the performance of their duties, the supervisor should be notified immediately. The guard will write a detailed account of the event, expressly including all data as well as narratives of the event. The security manager should facilitate this process by assuring that there is a uniform report that officers use.

USE OF EQUIPMENT

A directive should govern the use of equipment for official purposes only and only in the line of duty, e.g.:

All property issued to the employee is for official use only and is intended to be used only in the line of duty when circumstances warrant their use.

UNIFORM REQUIREMENTS

A directive should govern the uniform and what may or may not be displayed on the uniform, e.g.:

The uniform is for duty use only. Employees are permitted to wear the uniform to, from, and while on duty. Only authorized badges and insignias may be displayed on the uniform.

REPORT WRITING

A policy on reports should include when a report should be written and what should be included in the report, e.g.:

Security reports must be completed when:
    There is a violation of a security order or safety rule
    Someone reports a crime to security
    An alarm is activated

CONCLUSION

A policy is the organization's guiding or governing principle. It has been the intent of this chapter to assure that a general guideline is set forth with parameters for decision making along with the authority for implementation. Policy development should be a planned process requiring due diligence in thought because a policy becomes a strong principle of the organization. Policy justification should reflect problem solving while demonstrating persuasive reasoning, clarity, and coherence.

EXERCISES

It has come to the attention of security management that your organization is having issues with employees leaving work early and having another employee clock out for them later. Write a policy that will deal with this situation. How will the policy be enforced? Is there an incentive (including negative incentives)? Is there support for the policy from the workforce? Is there some overriding need for the policy? The policy should be consistent with the organization's mission, culture, strategy, and vision. It should not overlap or contradict with other policies and procedures.

6 Staffing the Response Force

Security response forces are part of a comprehensive security program (National Advisory Commission, 1976). Facilities that operate without a trained response force must rely on local law enforcement to respond to an event requested by a management group that may or may not have any knowledge or foresight of security procedures or needs. A receptionist or clerical employee assumes the task of screening visitors and also initiates the first response to a hazardous event. Distractions from their normal tasks, and the fact that security is not their forte, means relying on such for security is inadequate, if nonexistent. The decision to have an onsite response force takes into consideration the volume of employee and visitor traffic, the value or sensitivity of the assets and resources being protected, local crime rates, the availability of local law enforcement to respond and their response time, and other security measures in place (see Chapter 7, Physical Controls).

A TRAINED RESPONSE FORCE

A security force is a deterrent to criminal activity. Uniformed officers send a message that the site is guarded and an immediate response is available. Officers can supervise access control. They can secure exterior areas by having a guard house located away from buildings to verify authority for persons to be on the premises. Employees and visitors can be screened and deliveries checked before moving on to buildings. At building entrances, guards verify the identity of employees whose access control badge may be misread by electronic systems, as well as operating screening equipment, such as magnetometers and x-ray equipment. Extremely secure sites may have explosives sniffers and the capability to x-ray vehicles, along with vehicle barricades.

Visitor badges can be issued and controlled as well as have officers escort key visitors onsite. The force works with Human Resources to issue identification badges or cards and entering this information into electronic access control. They input new employees into the system and delete and reissue lost badges. In addition to badge and ID control, security monitors access to facilities.

Stationary posts are needed for access control, monitoring of intrusion detection systems, which include alarms and closed circuit television (CCTV) monitors. While not necessarily a security function, security is used to identify and monitor hazardous conditions, such as a chemical spill or other accident, such as acts of notice.

Security patrols can be used to respond to alarms, fire, or other security or safety concerns. Patrols can be dispatched to escort fire or emergency personnel to a location. However, security barriers may need moving to allow for equipment to respond. In one example, for security purposes, a site had closed off the drive adjacent to the building where the fire was occurring. Portable Jersey barriers (modular concrete barriers) were placed at the ends of the drive to prevent vehicles from entering. In this fire event, the security personnel brought in a forklift to move the barriers so that the fire equipment could have closer access to the building.

Patrols also monitor open doors and unattended equipment. During patrol, guards can identify and record security concerns, such as broken windows or locks, or safety concerns, such as loose steps, hallways lights out, or emergency exits blocked. More intensive patrols can check for open files, computers that are left on, and other issues mandated by the security policy.

Patrols can be monitored by Guard Tour Wand Systems. This system records when a guard has patrolled the area by having an electronic wand touch a memory button fixed to a wall or door at the facility. Placing memory buttons in strategic places assures that the patrol passes by and views specific areas on their tour. A wand is used to record this activity. Managers download the data to check that patrols are being performed correctly and on schedule. An example would be a site with multiple buildings and perimeter fencing. Policy may dictate that security would patrol the site hourly. Memory buttons may be placed in and outside of buildings, gates, and fences to monitor patrol progress.

Determining the manpower requirements to operate a private response force requires a detailed analysis of the security response requirements. Initial considerations should include how many stationary and patrol posts are needed and whether the post will be operational 24/7 or just operational during certain times. For example, an employee entrance may require one guard present 24 hours a day, but, at high demand times, such as shift changes, other guards may be needed. An employee gate may be open during high demand times, then closed during the rest of the day. A patrol guard can work the gate for the limited period it is open, then go back on patrol. Delivery gates may be opened for a full day, but not at night, necessitating a guard for a full shift. The security office, with the alarm and CCTV monitors, can be strategically placed so the guard can monitor the controls and assist in other functions during slack hours. The security manager should consider that, when applying one person to do multiple jobs, it decreases the efficiency of that person. A guard called to monitor cameras while manning the front desk will have his/her attention diverted when people need assistance, during which time, the monitors will be unattended.

A question for your company may be: How many patrols are required? Security policy or deterrent requirements help determine the number of patrols needed. A large facility requiring hourly building checks will need more patrols than a small site with few buildings. Patrols can be adjusted for time just as stationary posts. For example, a 24-hour patrol post may fill in at a gate during a 2-hour shift change time. Patrols also provide staff break relief for stationary posts. A supervisor generally assists on patrol and reliefs as necessary. Because of unanticipated events, the supervisor should not be assigned to a regular patrol. Other questions arise. How will the patrols operate? Will the guard be on foot, in a vehicle, or both? Are there specialty vehicles that guards will use, such as a golf cart or bicycle? How many people do you need to fill these posts? One post, operational 24/7, requires 4.2 persons. The security manager must consider this as well as scheduling to justify the manpower needed to fill the posts. While most work will be done by full-time personnel, part-time people may be used during certain times or days. For example, a site calling for one guard may employ four guards, one of whom is the supervisor, working four 40-hour shifts, and one part-time guard working one 8-hour shift. Days off, sick days, and vacation time must be considered. In extreme circumstances, guards may be required to work overtime. Scheduling overtime increases manpower costs. Employee efficiency decreases with time, so managers need to be aware of excess overtime. Instead of having a guard work a double-shift to cover for a sick co-worker, it may be advisable to have one guard stay over for one half of the sick co-worker's shift, and another guard come in early to work the other half of the shift. Another consideration is time for training or administrative matters.

Shift changes for the guard force should not conflict with the work force. If the day shift arrives at 7:30 a.m. for an 8 start time, you don't want a shift change at 8. You may decide that a 6 a.m. to 2 p.m. shift is best so that you have fresh guards on as people arrive for work and there is no confusion as to briefing the oncoming shift of special needs and circumstances while the urgency of personnel ingress is happening. A fresh guard should be rested and ready to work. He/she will not be in such a hurry to "get this over with and go home."

What equipment will the guards need? A guard force should be in uniform, meaning anyone can identify a guard for crime deterrent or in the event of an emergency. There are normal needs for a guard force, such as uniform shirt and pants, cold weather coats, and raingear. A hat is customary, but not necessary. Most uniforms are military/police style to present a show of authority, but some response forces use a less aggressive-looking uniform, such as slacks and a blazer. Badges also display authority. The joke about security guards often refers to "sew-on badges." Metal badges have a show of authority and gain more respect for the guard.

Safety gear should be provided. If a facility has hazardous chemicals, do the guards have protective equipment in the event of a spill or leak? Steel toe shoes or hard hats may be required in certain areas. Radios, tour wands, and flashlights may fill out the complement of equipment needed for duty. If vehicles are used, the security manager should make sure that the guard is capable of operating the vehicle and is a properly licensed driver.

While the guards may be required to provide some of their own equipment, policy should dictate uniformity. For example, most guards will wear black shoes or boots. The type of shoe may be left to the user with limited parameters (e.g., shoes must be black) or severely restricted, specifying the type of shoe allowed (black patent leather with steel toes).

The organization should develop standard operating procedures for the guard force to follow. The duties and responsibilities of the guard force, as well as individual posts, should be identified and published in a procedures manual. (See Chapter 5, Policy and Procedures, for more information.)

The decision to provide defensive equipment should be determined by policy. Consideration as to nonlethal force should be given the same care and concern as deadly force. If the decision is made to allow weapons, training should be provided to assure proper use. If the decision is made not to allow weapons, policy and procedures should dictate the protocol for the protection of the guard.

Now that you have determined the personnel and resource needs for your facility, you may choose to have either a proprietary or contract guard force. A proprietary force is owned and managed by your organization. They are your guards, and your organization is responsible for hiring, salary and benefits, training, and equipping of the guards. Liability rests solely on your organization for intentional or negligent acts of the individual guards. A contract guard force is managed by a contract company (often a company specializing in guards) for a specified contract price. They hire, provide salary and benefits, train, and equip the guards. Your organization will have someone who will be the liaison with the contract agency. In some matters, the guard company may be liable for intentional or negligent acts of a guard. The liaison will be responsible for assuring the contract agency is performing according to the specifications in the contract and meeting with agency representatives for routine performance reviews.

Guards may be represented by a union and operate under a collective bargaining agreement with management. If you have a proprietary force, that is you. The security manager must work with Human Resources and legal counsel to assure that discipline, scheduling, and promotions are consistent with the agreement.

Regardless of whether you have a proprietary or contract force, a job assessment will identify the needs for the guard force as well as begin the process for writing a job description. To write a valid job description, begin with listing all of the tasks a guard will perform, such as answering the telephone, writing reports, watching monitors, operating machinery. Any movements should be documented to justify requirements of the Americans with Disabilities Act (ADA). The ADA protects against discrimination of a person who can perform the job functions, with or without reasonable accommodation. For example, list tasks such as standing, walking, negotiating stairs, climbing, crawling, reaching, etc. Vision and hearing requirements, as well as the ability to function specific machinery, should be included in the list. As you screen applications for employment, those who cannot perform the functions, with or without reasonable accommodation, can be excluded from consideration. Is it important to write a thorough job description so potential applicants will know the requirements and apply for jobs for which they are qualified.

The job description is also key to training. Training assures that the employee can perform job tasks proficiently. It minimizes civil damages in the event of a lawsuit. Training may be in the classroom, online, on-the-job (OJT), or a combination of methods. Contract agencies provide a level of training for uniformity of their guard force. A proprietary guard force may be trained onsite or sent to a school that provides a basic level of training. Private corporations and security organizations have ready-made materials for use in the general training of your guard force. Depending on specific needs, other prepared training programs may be helpful and even necessary. For example, a guard at a casino where alcohol is served may have required training from the state gaming commission as well as the alcohol beverage commission.

Guards should be aware of the policies and procedures, company policies, operational needs, and priorities. Orientation should include a copy of the policy manual. After a reasonable time, have the guard sign a form indicating they have read the policy manual.

Training should include general duties applicable to the guard force as well as specific duties for each post the individual guard may work. Administrative matters, such as paydays, insurance, and time off can be disseminated at this time. Additional training should be conducted based on the frequency or criticality of a specific function. For example, reports that are completed daily have a high frequency, so training on report writing would be a reasonable training need. Use of force may not be frequent, but highly critical as the improper use of force can lead to injury and possible lawsuits. Training the parameters of force will assure proper conduct of the guards and mitigate damages if and when force is used.

Other training needs would entail patrol or station procedures, customer service, telephone communications, and other skills not necessarily related specifically to the trade, but for events that could impact the organization. For example, when a guard answers the telephone or gives instructions to visitors, he/she must sound (and appear) professional. Answering the phone with: "Yea, what can I do for you?" is quite informal and may present a negative impression of the organization. Proper procedures should be identified and taught to all personnel.

Training should include all equipment the guard will use. This includes vehicles, patrol wands, weapons, and safety equipment. Guards are often the first responder to a fire, safety, or medical emergency. Knowledge and use of fire extinguishers and other firefighting equipment is essential. The guard should know when and how to deploy and use safety equipment, such as personal protective equipment, a fall protection safety harness, or safety barriers. Training should include OSHA (Occupational Safety and Health Administration) rules that apply to the facility and specifically to activities the guard may encounter and areas where the guard must patrol. Guards should be familiar with any first aid equipment and automated external defibrillators (AED), if available. They should have a working knowledge of emergency evacuation and shelter-in-place (taking immediate shelter) procedures. The guard force should always participate in all drills and exercises. Most organizations expect the guard force to coordinate emergency activities.

EXERCISES

Explain why an organization would need a trained response force versus relying on public law enforcement. What are the conditions in which organization would utilize a response force?

REFERENCE

National Advisory Commission on Criminal Justice Standards and Goals (NAC-CJSG). 1976. *Private security: Report of the task force on private security.* Washington, D.C.: U.S. Department of Justice, Law Enforcement Assistance Administration.

7 Physical Controls

Physical controls are the tools and equipment one uses to secure assets. A physical security survey and assessment is used to determine the physical controls necessary to secure a facility and protect assets. The survey begins with an asset and risk assessment to determine your security needs. Identifying the assets to be protected helps justify the expense of the physical controls. You wouldn't spend thousands of dollars in security hardware to protect an asset costing pennies. If one area of your facility houses high-risk assets, you wouldn't necessarily provide the same level of protection for the entire facility. High-risk assets are any assets that are expensive or hard to replace. It could be anything from diamonds to information.

The risk assessment looks at who would want your assets and their level of knowledge and expertise that you must protect against. The thief wanting to steal the diamond may be low-tech and utilize crude burglary techniques with only a plan for a quick getaway. The professional engaging in industrial espionage will have access to high-tech gadgetry and extensive knowledge of physical control systems to ply his/her trade.

With the results of your survey in hand, you begin the process of providing physical controls to protect your assets. This chapter is an introduction to physical controls and only touches on basic information of each tool. Security managers should consider taking additional courses or training on physical security controls to gain a working knowledge of the various systems. Physical controls should be planned with the objective of preventing entry through force and stealth of unauthorized persons while allowing entry and safety to authorized persons. The controls should delay unlawful entry until security forces arrive. Another consideration is attacks solely for the purpose of destruction, such as in a terrorist act, or sabotage from, for example, disgruntled employees, former or current.

THE SURVEY

A physical security survey begins with presurvey planning. Previous surveys are a good starting point. If a prior survey doesn't exist, begin by getting a map of the site showing buildings, utilities, and roads around the site. Determine if there have been changes that are not noted on blueprints or site plans. Identify future changes that may impact security. Obtain crime statistics from the local police to determine possible threats or concerns for security and safety of personnel.

Next, identify the assets onsite and the cost of loss, compromise, and ease of replacement. The value of a specific item that is stolen or destroyed may have a severe economic impact, not only by the loss of the item, but the ability of the item to create revenue. An oven for a restaurant may be easily replaced, but it could take several days to get a commercial one delivered and the restaurant cannot operate until a new one is installed. However, the "secret" recipes relied on by the restaurant can bring everything to a halt in the kitchen as well as the threat of competitors getting your trade secret if they are divulged.

Once the assets are identified, review the present security hardware and policies to determine if they are sufficient to deter or prevent the loss or compromise of the key assets. Threat decomposition looks at each potential threat and the ability of that threat to defeat the security measures in place. For example, a truck yard housing trailers loaded with your product is susceptible to theft and/or vandalism. Your threat decomposition would include identifying potential persons who may be a threat (auto thief or gang) and their ability to accomplish their threat (theft or vandalism).

SITE VISIT AND ADJOINING PROPERTY: NEIGHBORHOOD

Layers of Security

All facilities have some level of security, however slight. The 24-hour restaurant with no locks on the doors may have a safe and a lock on an interior office door where the safe is located. Security managers usually take the facility they have surveyed and examine current security and how specific upgrades will provide adequate protection for assets. However, not every facility needs large-scale protection. The tendency of security managers is to see physical controls as something high tech and intimidating. We forget that the simple key in the knob lock, inadequate for most applications, is still physical security.

The security manager should be part of the team when building a new facility or remodeling or adding on to an existing facility. Crime Prevention Through Environmental Design (CPTED) allows architects and security managers to build the new facility and remodel existing facilities using techniques with protection built in. This is helpful when a facility is located in an area that forbids fortress-looking facilities. Organizations don't want their facilities to look like a fort; security can be built into an aesthetic design without looking like security. For example, instead of unsightly bollards (short, thick posts) lining the street side of your facility, heavy-duty planters designed to stop vehicles can protect your facility and, at the same time, be filled with colorful, attractive plants.

Perimeter

Begin by examining the facility construction. The building wall is usually the first layer of defense. A concrete wall protects more than a metal-sided building. Interior walls of sheetrock can be punched through with hand tools, affording little security once a perpetrator is inside your facility. Doors and windows should be visible with lighting so the burglar will not be concealed while he/she works. Most unlawful entries occur through the first floor with the front door being the preferred method of entry. This does not reduce the need to secure windows and rear entrances—or the roof. Rear doors and windows are especially vulnerable as they are often out of public view. All exterior and some interior doors should be secured with locks designed to withstand forcible assaults or picking. Emergency regulations often require exit doors to open outward, exposing the hinges. Commercial doors are available with recessed hinges to thwart such attacks. Doors with exposed hinges should be pinned to prevent removal of the hinges and entry from opening the opposite side of the door. On interior doors, it doesn't do any good to have a strong, sturdy door if all the perpetrator has to do is break through a thin wall.

Windows can be coated with a safety glazing that reduces shattering. The glazing holds the glass together on impact, slowing down entry. The glazing often has aesthetic properties like reducing glare and energy efficiency. Window locks should function. Locks should be tested to determine the difficulty in manipulating them open. Weak locks should be replaced or reinforced. In some applications, a window may serve as an emergency exit.

Some facilities extend their perimeter by providing landscaping between the building and parking lot, which provides aesthetics and some protection. Landscaping can deter various attacks. Trees, planters, uneven landscaping, and soft ground are examples of ways to deter someone driving over the surface. Lighting and shrubbery also may deter pedestrian traffic. Someone walking on the grass is an indication he or she doesn't belong, providing more time for the response force to be notified and to arrive on the scene. Other sites may have the property enclosed in a fence. Exterior alarms, closed circuit television (CCTV), and security patrols help detect perpetrators before they reach the building.

Fencing

Fencing can be standard security materials or composed of attractive, yet useful, materials. A standard security fence is 11 gauge chain links with a maximum of 2-inch mesh openings. The fence should be 7 feet tall above ground with 1 foot buried. Contrasting gravel on each side of the fence helps maintenance through good drainage. The fence should be topped with an outrigger of barbed wire. Some locations have a Y post to position barbed wire facing out and in toward the facility. Additional measures can be concertina or razor wire arranged in the Y post.

Security fencing should help prevent a vehicle attack. Bollards or security-designed planters work to slow and stop a moving car or truck. TigerTraps (collapsible sidewalks and planting system) are areas designed to withstand light pedestrian traffic, but would collapse under the pressure of a vehicle. A common and effective means of supplanting a security fence with an antiram system is to secure a heavy cable at vehicle height. The old adage, a chain is only as strong as its weakest link, applies to fencing. When applying antiram hardware, an upgrade of posts will enhance the ability of stopping a vehicle attack. Security fencing is being designed with antiram capabilities.

Aesthetics and local codes may prevent security fencing from being used at a given site. Fencing materials, such as wrought iron, brick, and combinations of materials, can deter and delay entry while being attractive. High security applications may use smaller mesh, thicker wire, and fencing designed to discourage climbing or being attacked with hand tools.

With the need for security in urban areas, sites do not have the luxury of space. If fencing is used, keep parking away from the fence. Also keep trees and other objects clear of the fence so as to provide a clear view of the fence line and not to give the intruder tools with which to climb.

Gates allow entry through perimeter fences. You may need a full gate for remote areas, but common entrances may be secured with bollards and a guard. Full gates should have an antiram system that links with the fences and their antiram system. Entrances with a barrier capable of emergency deployment should keep the barrier up and lowered when a vehicle is allowed access. Emergency deployment of the barriers should be used as a last resort, considering that the use of force may be deadly and the systems are not designed for continuous rapid deployment.

Access Control: Keyed Locks

A secured site must allow people into the facility to conduct business. Access control systems are designed to allow access to the right people while denying access to the wrong people. Simple systems use locks with mechanical keys. Users are issued a key that will open a door or series of doors. Key control is vital as anyone who possesses the key can open the door. Management must determine which doors need locks and who needs keys to those locks. The best systems use commercial-grade locking systems designed to withstand picking and copying. These types of systems usually include removable cores for replacing and rekeying locks. A lock should be rekeyed when a person leaves the organization or when a key is lost. Commercial systems are designed to allow for such changes with a minimum of cost. Although lock and key companies work to make this easy, it is a hassle and few companies follow through with changing locks and keys. Persons issued keys should be instructed to keep the key in their control and not lend it out or leave it where it can be copied. A more effective system is electronic access control.

Electronic access control can use a system where an employee enters a combination to open a lock. This type of system utilizes unique knowledge: the combination. As with the key, anyone who possesses the combination can open the door. A more intelligent system uses a fob or card that activates an electronic reader attached to the doors. Cards/fobs are issued to individual users and can be programmed to allow entry only at specific times. They also track when the user has accessed the lock. This creates a record of who opened the door and when it was opened. If a card/fob is lost, or the employee's access to an area is terminated, the card/fob can be decommissioned, making it useless for entry.

More secure systems use a combination of unique possession and unique knowledge. For example, a door may require a key and a combination to open. Adding multiple layers to the system enhances security. A person may obtain your key or card, but wouldn't know the combination. Extremely secure systems use retina or iris scans, hand geometry, and fingerprints to secure entry.

Lighting

Security lighting should be designed to enhance observation as a means of identifying perpetrators and deterring persons from unlawful entry by limiting dark areas where a person could be concealed. Continuous lighting is the most common, with fixed light sources illuminating a given area during dark or low-light conditions. These systems operate on a timer or photocell to turn lights on when needed and off when not needed. Standards for lighting vary among industries. The Illuminating Engineers Society has a minimum standard for lighting applications that include security lighting. Other applications may have a higher standard. Lighting type and use should assure safety and security, e.g., parking lots should have a light level to afford safety while walking and while illuminating dark areas where an attacker could hide. Another consideration for light sources is the type of lighting when CCTV is used. A color camera CCTV would require a light source with a broad color rendition to be effective.

Security lighting is rated by its efficiency and color rendition. Energy costs for lighting vary between sources of lighting. A lumen is the amount of light emitted by a source. A typical incandescent light emits 28 lumens per watt (the least efficient) while a low pressure sodium light emits 200 lumens per watt (most efficient). Operational costs would dictate the use of low pressure sodium in energy savings alone.

The lifetime of lights must be considered as well. An incandescent light bulb normally lasts 1,000 hours while low pressure sodium lights last 20,000 hours. Longer life means less replacement. Another factor is color rendering. The incandescent bulb renders light in a spectrum with a color rendering index (CRI) of 100. This means there is a broad spectrum of light emitting from the incandescent bulb with colors appearing similar to what you see in daylight. Low pressure sodium is −44, which means it emits a low spectrum of light with color limited to yellows and grays. Flesh color appears gray and colors dull. While this type of lighting is cost efficient, it would not be useful for color camera CCTV operations or at a location where color-coded materials are used.

Other lights for security applications are high-pressure sodium, which has a good CRI and lifespan; metal halide, which closely replicates daylight, even enhancing colors (car lots use metal halide); and LED, recently entering the security lighting market. LEDs emit 160–200 lumens per watt and can last up to 50,000 hours with excellent CRI. The security manager may consider a combination of light sources: low pressure sodium in parking lots with metal halide or LED on wall packs and at entrances and where color CCTV is in operation.

Intrusion Detection Systems

Intrusion detection systems (IDS) alert the response force when there is a breach of a security zone. A basic system includes a sensor, an annunciator, and a means of transmitting information from the sensor to the annunciator. An alarm does nothing more than tell you there has been a breach. A response force completes the system. The alarm notifies you of the breach, and your perimeter and locks should be designed to secure your assets until the response force arrives. The earlier you detect an unlawful entry, the faster you can have the response force reacting to the attack. Common systems use motion detection sensors that transmit to an audible alarm or central station using radio frequencies.

There are interior and exterior systems. Most people are familiar with the interior systems. Sensors are placed at entrances or approaches to sensitive areas. Doors and windows may have magnetic switches that detect when a door or window is opened. Windows may have glass break sensors that detect acoustically or through shockwaves the breaking of glass. Wall sensors may detect vibration or sound.

Volumetric sensors detect motion by the disruption of an electric field or a light beam. Beam sensors direct a light path from a sending unit to a receiving unit, such as a garage door safety switch. Beams may be directed with mirrors and filters to create an invisible matrix of light that, when broken, sounds an alarm. An infrared system, commonly known as passive infrared (PIR) sends out a signal and can detect changes in temperature between narrowly designed zones. When a person attempts to move through the zone, the heat difference from their body temperature and the room temperature disrupts the fields and signals an alarm.

Exterior systems consist of fence sensors, in-ground sensors, and volumetric sensors. Fence sensors send an alarm when there is a disturbance on the fence. A shake detector is installed on a fence and uses free-floating contacts that move when the fence is moved, typically by climbing, cutting, or lifting. However, strong winds or animals striking the fence may set off an alarm. Taut wire senses a change in tension on the fence. Buried line sensors are similar to the PIR sensors in that there is an alarm when there is a change in the electric field. A buried line creates a field of energy that is disrupted when someone passes through the field. Microwaves act in a similar manner as the beam of light by detecting when there is a disturbance in the microwave beam. Seismic detectors sense vibrations that may indicate an intrusion.

Fiber optics are used in IDS applications by two means: speckle pattern and interferometry. In the speckle pattern, light sent through the fiber optic cable appears on a monitor as a pattern of light and dark. The pattern changes when the field is disturbed, indicating a break-in. Interferometry uses a beam splitter to send signals in different directions at different wavelengths. A disturbance interferes with the time it takes two beams to meet, thus signaling an alarm. Video motion detection is similar to the fiber optic in that a CCTV monitor can be programmed to alarm when there is a change in pixels on the screen. Adding to this are new biometric systems that are accurate in detecting anomalies and individual characteristics. The system can be set to alert the response force if a vehicle moves into a specified area or it can be programmed to identify normal users and alert the response force if an imposter is trying to enter a secure area.

CCTV

Closed circuit television is a means of extending your security surveillance. A system uses cameras and monitors for a broader coverage than security patrols. Cameras can be placed anywhere. Usually, cameras are located in areas requiring additional security or in an area difficult to patrol. Remote locations also benefit from CCTV coverage. Software can make your CCTV event-driven, monitoring changes in pixel color to alert for possible intruders. Technology can capture images and compare them to a database for access control. In addition to monitoring controlled areas and assessing a developing situation, digital technology can record events and is accessible over the Internet. Internet transmission and the low cost for system hardware make CCTV an inexpensive addition to your physical security profile.

Cameras can display black and white or color images. Black and white usually has sharper contrast, allowing for better resolution especially in low-light applications. Color is good when you need to distinguish items, such as a security badge or uniform color. Lighting is important with color cameras as enhancement from lighting may give off false colors. Cameras may be fixed, meaning they are focused on one area and cannot be moved. Other cameras are PTZ, or pan/tilt/zoom, and can be moved to observe a specific area. PTZ cameras can be programmed to turn and focus on a specific area during an alarm.

Cameras should be focused according to what you are trying to accomplish. If your objective is to detect, focus can be broad because you are only trying to see a person breaching the area covered by the camera; a closer focus, so you can later identify an intruder, shows less area. For example, an exterior camera monitoring 500 feet of fence line may only need to detect an object the size of a person entering the area while a camera in a bank needs to be able to identify a person's face.

Cameras should be positioned to protect them against vandalism. A popular housing is darkened so intruders cannot see where a camera is aimed. The housing is also inexpensive to replace should an intruder attempt to disable the camera with paint or other substance. Exterior camera housings should protect against the elements as well.

Monitors come in multiple sizes and can be programmed to put a number of frames on one monitor. A guard watching one or several monitors will have his or her attention diverted with multiple monitors. Software technology to warn of a disturbance helps assure that the guard will observe an event so he/she can dispatch security patrols to the scene.

EXERCISES

1. Expand on the idea related to layered security in that layers of security concept arises from the desire to cover for the failings of each component by combining components into a single, comprehensive strategy, the whole of which is greater than the sum of its parts.
2. Give an example of five layers of physical security surrounding petty cash that is stored in the office of a school.

8 Leadership

INTRODUCTION

Leadership motivates people. People design, develop, and implement work processes, and leadership motivates them to follow and achieve these process goals. Management controls systems by measuring the processes to achieve an efficient and effective flow of resources for productivity. Like management, leadership can be learned. Most anyone has the potential to be a leader. Training and assessments generally focus on the potential leader's capabilities by examining three core skills: (1) conceptual ability to create a vision and mission of the organization, (2) technical skill in work processes, and (3) human skill in relationship building.

Vision is key to leadership. Good leaders are compassionate when it comes to their visions. They can have a zealous focus on organizational and personal goals. Executive-level leaders focus on vision and on moving the organization forward. They inspire people by creating an ownership in the vision, and with this shared vision, enable people to act. Good leaders are willing to challenge the status quo and to innovate, which empowers others to also achieve the organizational goals. Technical skill deals with knowledge of a specific work or activity. First-line leaders generally have more skill and are focused on mission accomplishment. As the leader rises through the organization, less technical skill and more organizational vision is needed. Human skill means working with people at all levels. Good leaders should lead the way by living the values they expect of all in their group.

Traits of a good leader include intelligence, self-confidence, determination, sociability, and integrity. Sometimes difficult to measure, individuals can enhance these traits and skills. Intelligence traits include strong communication skills (verbal and written) and the ability to properly notice events and direct an effective response to events. Self-assurance and positive self-esteem gauge the leader's self-confidence. Determination measures the leader's drive to accomplish the mission and tasks. Sociability implies the leader must possess good interpersonal skills that create cooperative relationships. Integrity is the most important characteristic of a leader, because leaders without integrity lead only until their followers lose trust in them. Leaders with integrity inspire confidence in those under them. No one cares how much you know until he/she knows how much you care.

Leaders must be genuine and real. Their actions must be congruent with their values. They build relationships with others. Social skills are essential for influence, which is the major characteristic of leadership. The ethics of leaders impact the ethical climate of their organizations by taking into account the interests of the organizational culture. Ethical theory provides a set of rules or principles that guide leaders in making decisions. Ethics is central to leadership because of influence. Ethical leaders respect the values and opinions of others. They listen to subordinates and are open to opposing viewpoints. Conduct and character are two domains of ethical theory. Ethical leaders are open, tell the truth, and hold themselves and their followers accountable. They treat all equally and assure that all treatment is fair.

Good leaders have influence, which may come from the position rather than the individual. Weak leaders rely on the authority of their position to compensate for the lack of personal leadership traits and skills. Some resort to fear to influence others in a negative sense. Highly effective leaders influence with personal power. Personal power can be developed by identifying and cultivating key characteristics that make for the ability to influence. Influence comes from integrity; leaders must have integrity, and must be trustworthy at all times. Virtue-based theories focus on the leader's character. Practicing good values lead to increased virtuousness, which lead to a more effective leader.

Charismatic leaders are confident and have a dominant personality. Strong moral values and a desire to influence imparts a strong role model and trust in the leader. Charismatic leaders express confidence in their followers and articulate their goals and vision in a way that is accepted..

TYPES OF LEADERSHIP

TRANSFORMATIONAL LEADERSHIP

More than just charisma is required for transformational leadership. As leadership is about influencing others, transformational leadership focuses on the morality of the leader and in directing activities in a positive way. Transformational leaders must have a clear vision of the future and communicate that vision in an understandable language that also is energizing. They are the social engineers of the organization, and they shape the culture of the group. They create trust through transparency, i.e., making their positions known to all and concentrating on their strengths rather than dwelling on their weaknesses. Transformational leaders motivate followers to reach their full potential.

Transformational leadership emphasizes influence and inspirational motivation. The individual is connected to the goal of the leader through emotion and logic (Northouse, 2006). Idealized influence describes the role-modeling process for followers. Strong moral and ethical standards cause followers to imitate them. Inspirational motivation occurs when the leader clearly communicates the vision and sets high expectations of the followers. Intellectual stimulation encourages creativity and innovation. Followers learn to challenge their own beliefs and to recognize personal bias that negatively impacts mission goals. Individualized consideration focuses on the follower with the leader acting as coach, helping the follower achieve his/her full potential.

In servant, or follower-centered, leadership (coined by Robert Greenleaf in *The Servant as Leader*, 1970), leaders strive to serve others. They help others achieve their goals, thereby nurturing the follower with the vision of the leader. Their decisions consider the impact on the welfare of all. The servant leader reaches beyond his/her aspirations placing the goals of the group above personal goals. The leader makes sure the followers have the resources necessary to accomplish their goals.

Within the various attributes mentioned, leaders' skills are directed toward individual attributes, competencies, and leadership outcomes. Individual attributes of the leader are cognitive ability, motivation, and personality. Competency examines subject matter knowledge, problem solving, and social judgment. Leadership outcomes observe the ability in problem solving directed toward performance criteria.

Career experiences and environmental factors influence individual skills. Career experience looks at hands-on knowledge, the ability to handle challenging assignments, the ability to mentor new employees, and provide appropriate training for subordinates. Environmental influences are factors outside the person's control, such as outdated technology.

SITUATIONAL LEADERSHIP

Situational leadership is focused on directive and supportive behaviors. Depending on the "situation," the leader can change his/her style for an ideal fit to the follower's needs. The leader assesses the follower's competency and commitment to determine what style of direction is needed. The leader selects a style to match the follower's development level of behavior:

- high directive–low supportive
- High directive–high supportive
- High supportive–low directive
- Low supportive–low directive

Development levels also are broken into four areas

- Low competence–high commitment
- Some competence–low commitment
- Moderate/high competence–varied commitment
- High competence–high commitment

Take the example of a new employee. He/she may have low competence, but high commitment. The leadership behavior most appropriate would be high directive. The new employee needs direction to help him/her become more competent. As the employee develops competence, the leader moves to coaching, with the leader continuing to give direction and adding high support to the process. Eventually, the employee develops high competence and the leader can delegate the tasks totally to the employee.

Situational leadership has survived the test of time and proved to be an effective means of training leaders within an organization. The system works by having the leader apply the proper style to the individual follower. It is easy to understand and apply. A concern of the system is that people don't always fit nicely into one of the categories. With practice, the leader can learn quickly how to adapt between behaviors.

CONTINGENCY THEORY

Contingency theory is similar to situational leadership in that the leader's success depends on his/her ability to match his/her style with the situation. Contingency theory is supported with research that validates its ability to explain how effective leadership can be achieved. Like situational leadership, contingency theory is not tied to one "best" way of handling a situation, but rather teaches the leader to be flexible in his/her approach. Contingency theory sees leaders as either task motivated or relationship motivated. Three factors—leader–member relations, task structure, and position power—are key in using the contingency theory model. The emphasis of leader–member relations is how work teams get along. Task structure examines the specificity of instructions and task requirements. Position power is the authority of a leader to reward or punish followers (Northouse, 2006).

PATH-GOAL THEORY

The path–goal theory looks at the leader's style, the characteristics of the subordinate, and the work situation. Path–goal assumes that subordinates are motivated when they are capable of performing the work, believe their efforts will result in a specific outcome, and feel the reward for the work is worthwhile. Path–goal looks at the components of leader behaviors, subordinate characteristics, and task characteristics to determine the leader's impact on subordinate's motivation. The theory works by having the leader look at the needs of the subordinates, and fitting his/her style to fit that need.

Leader behaviors include: directive, supportive, participative, and achievement-oriented. Directive leaders set clear standards of performance as well as making the process rules clear. Supportive leaders are approachable and show concern of the subordinates' needs. Participative leaders invite subordinates to share in the decision making. Achievement-oriented leaders challenge subordinates to perform work at the highest levels possible. As with situational leadership, the leaders in path–goal theory will adapt their behavior to the needs of the subordinates (Northouse, 2006).

Subordinate characteristics include a need for affiliation, preference for structure, desire for self-control, and task ability. The leader's effectiveness depends on how he/she responds to subordinates with varying degrees of each characteristic.

Task characteristics provide motivation for subordinates through the task, formal authority, and the work group. The focus of the leader is to help subordinates through and around obstacles.

LEADER–MEMBER EXCHANGE THEORY

Leader–member exchange theory looks at leader–follower relations being aimed at individuals rather than at a group as a whole. While the workgroup has many members, the leader's success is focused on one follower at a time. Each subordinate has different characteristics that require the leader to tailor his/her approach to the individual. In a workgroup, subordinates may be seen as being part of the "in" group or part of the "out" group. Personality plays a role in this process as a bond between the leader and follower grows or wanes.

The leader–member exchange theory focuses attention on building relationships between the leader and followers, and between followers. The theory shows the importance of communication in leadership. The "in" group works better together and is usually more efficient. Members of the "out" group are less effective. Leaders need to be aware of this and strive to nurture those who are "out."

Team Leadership

The study of leadership examines work teams and explores ways to make teams more efficient for the task for which they were created. Sometimes, a work group is called a *team*, but the lack of purpose and community is no more than a collection of individuals. Teams are individuals who band together and jointly strive to achieve a common goal. The study of team leadership supposes that individuals other than the formal leader can perform critical leadership functions.

The leader is basically a medium for processing information. The team comes together for the purpose of a unified effort in processing necessary information. Characteristics of team effectiveness include a unified commitment, a results-driven structure, a collaborative climate, and external support. The team leadership model integrates mediation and monitoring concepts to a group rather than to an individual.

Effective team performance begins with a mental model of the situation. The team leader develops a mental picture of the situation and relates it to the team. The team then takes action to solve the problem. The team leader must decide to intervene in the team's efforts by asking three questions:

1. *Should I monitor the team or take action?* The leader watches for internal or external factors that need attention, then must decide if his/her input is needed.
2. *Should I intervene to meet task or relational needs?* When the answer to no. 1 is to take action, the leader focuses on the need to be addressed: Is there a problem with the task, a problem with group relations, or both?
3. *Should I intervene internally or externally?* Internal support is needed at times when you must clarify goals, facilitate decision making, and emphasizing standards of excellence. External support is when you must work in the organization to give the team credibility. External support may be gathering information from other functions so the team can function efficiently.

Psychodynamic Approach

According to the psychodynamic approach, learned patterns of family dynamics influence leadership. The insight into one's own personality is beneficial. This approach is criticized because it is based on the psychology of the abnormal and focuses primarily on personalities of leaders and followers. As such, it does not lend itself to traditional training.

Transactional analysis maintains that there are three ego states: parent, child, and adult. We shift in and out of the states in relationships. Transactions are complementary when they match the ego state of the other party, e.g., an adult ego state is matched by an adult ego. Dr. Sigmund Freud identified three personality types as erotic (the desire to be loved), obsessive (requiring order and stability), and narcissistic (takes pride in personal accomplishments). A fourth personality type was added by Erich Fromm: marketer (adaptable to change). Carl Jung expressed four dimensions in assessing personality: Extraversion versus introversion, sensing versus intuiting, thinking versus feeling, and judging versus perceiving. These can be combined into 16 combinations. Leaders should learn their own style; understanding your style will help you develop your skills in being effective.

LEADERSHIP AND DECISION MAKING

Leaders are expected to make decisions that further organizational goals. Leaders need the right information to make effective decisions. The leader needs to know what is going on. The trust gained in leadership encourages others to provide the leader with up-to-date information, the right information, and sufficient information to make an informed decision about the work process. The leader can assess the employees to identify their organizational concerns. Leaders should identify processes that need improvement and gather data to support a change or status quo. Some issues may be evident but misleading, such as groupthink. The data will show if an issue is truly cause for concern and help prioritize multiple issues.

Groupthink is a problem where one or more people make a comment that suddenly everyone believes on face value without the support of data. Meetings to gather data and discuss options should be open to involved stakeholders with all present able to voice concerns, with the understanding that each concern should be supported with information before making a change. Brainstorming is a process that allows people to come together in such an environment. Participants' statements are not challenged or supported, but accepted as a means of identifying the root cause of a problem. In fact, proper brainstorming techniques encourage others to build on previous comments. Another method similar to brainstorming is the Delphi technique, where stakeholders submit anonymous suggestions to a facilitator or bulletin board for the purpose of eliciting new ideas or solutions to a problem. It is used as a consensus builder by having multiple rounds of questions to get the parties to agree.

An opposing method for decision making is constructive conflict. This method puts opposing views at odds and forces them to prove their position and often disprove or refute the opposing position. The leader acts as judge, weighing the evidence from one side against the other, asking questions to clarify points so as to come to a conclusion on the best course of action to take. Regardless of the method used, a decision must be made. All decisions should follow certain key values.

The decision must align with the mission and goals of the organization. Ask: What are we really trying to accomplish? It may be easy to make a decision with short-term consequences, but the leader must make the decision with the organizational mission and vision in mind. The leader also must align his/her decision with the core values of the organization. While a decision may further the mission, it may conflict with a stated core value. In the event that two core values are in opposition, the leaders must prioritize the values to assure the highest value takes precedence.

Next, the leader must evaluate the importance of the decision so he/she can allocate the appropriate resources and time to the decision. The leader must consider who will be impacted by the decision, both in and outside the organization. As the leader progresses with his/her decision, he/she must weigh all options. There may be obvious solutions, but new information and technology may prove other solutions more viable. This is where the brainstorming or Delphi methods may help identify stakeholder concerns and the buy-in for the decision from affected parties.

Now, make the decision. Not everyone, even with a consensus, will back you 100%. People fear the consequences of making a bad decision. If you make a bad decision, admit it, then make the right decision and move on.

EXERCISES

Expand on the following statement and apply it to a security organization: Manager and leader are two completely different roles, although we often use the terms interchangeably. Managers are facilitators of their team members' success, while a leader leads based on strengths, not titles. The best managers consistently allow different leaders to emerge and inspire their teammates to the next level.

REFERENCE

Greenleaf, R. 1970. The servant as leader. The Robert Greenleaf Center. Indianapolis.
Northouse, P. 2006. *Leadership: Theory and Practice*. Thousand Oaks, CA: Sage Publications

9 Risk Assessment for the Security Manager

For security managers, risk assessment is an essential component of their vocation. A risk assessment is the primary tool used to allocate security resources in the protection of organizational assets. The risk assessment looks at an organization's assets, accounting for both the criticality and vulnerability of the asset to determine the security investment. The comprehensive process of risk assessment is a process of critical thinking. The risk assessment is a methodical process of evaluating credible threats, identifying vulnerabilities, and assessing the potential consequences. The risk assessments are the basis for protecting organizational assets through the application of strategies and countermeasures to reduce the threat of, vulnerability to, and/or consequences from a loss event.

OVERVIEW OF RISK ASSESSMENT

The risk assessment is a primary component of the risk management process. The objective of the risk assessment is to identify risks to organizational assets so as to propose an achievable level of protection that is commensurate with the level of risk, without exceeding that level of risk, and be cost effective as well. Risk is a function of the values of threat, vulnerability, and the collateral damage via loss occurrences. The objective of risk management is to create a level of protection that mitigates vulnerabilities to threats and their potential consequences, thereby reducing risk to an acceptable level. Ideally, all risk would be eliminated. However, in practicality, the elimination of risk is not feasible. (DHS, 2008)

The security manager, in consultation with organizational leadership, plays a critical role in the security decision-making processes. To make an informed, risk-based decision regarding the mitigation or the acceptance of risk, collaboration between the security manager and organizational leadership is required. For any countermeasure that is recommended, the security manager must provide all information pertinent to the decision, including the nature of the threat, specific vulnerabilities, an understanding of the potential loss consequences, and the costs.

The process of risk assessment will not prevent adverse events from occurring; however, it enables the security manager to focus on those things that are likely to bring the greatest harm, and employ approaches that are likely to mitigate or prevent those incidents. Therefore, the risk assessment is not an end in and of itself, but rather part of organizational practices that include planning, preparedness, program evaluation, process improvement, and budget priority development. The value of a risk assessment is not in the determination of a particular course of action, but rather in the ability to distinguish between various choices within the larger context.

The assessment of risk should not necessitate a comprehensive risk assessment for an entire organization. There is not a specific risk assessment methodology that transcends all organizational environments. There are going to be differing risk assessments for differing organizational environments. However, all risk assessments should be rooted in a critical thinking methodology. Risk assessments should employ a reflective reasoning process based on an understanding of the organization and its environment. The assessment chosen should adhere to the fundamental principles of a sound risk assessment methodology. The methodology should follow the premise of critical thinking in that:

1. The methodology must be credible, and assess the threat, consequences, and vulnerability to specific loss events.
2. The methodology must be reproducible, and produce similar or identical results when applied by various security professionals.
3. The methodology must be defensible, and provide sufficient justification for deviation from the baseline.

A risk assessment is not going to be comprehensive to all organizational environments. Different environments require differing risk methodologies. In practice, various methodologies will provide varying outputs, from numbers and percentages to qualitative ratings. The security manager must determine what outputs from their respective methodologies correlate with the desired level of protection. In organizations where multiple risk assessments may be conducted, the security manager will need to evaluate the comprehensive findings and determine what countermeasure recommendations to implement from multiple risk assessments.

The risk assessment also should document existing levels of protection provided by the inherent qualities of risk mitigation, for example, geographical isolation or existing safeguards. Levels of risk determined for each undesirable event should be considered in terms of inherent mitigation by existing countermeasures that provide a commensurate level of protection, meaning the higher the risk, the higher the level of protection.

The Department of Homeland Security (DHS) defines a risk assessment as a product or process that collects information and assigns values to risks for the purpose of informing priorities, developing or comparing courses of action, and informing decision making, via the appraisal of the risks facing an entity, asset, system, network, geographic area, or other grouping. Simply, the risk assessment can be the resulting product created through assessment of the component parts of organizational risk. (DHS, 2011b)

A risk assessment is a decision-making tool. The risk assessment serves the security manager by identifying and analyzing threats and vulnerabilities, and provides the basis for recommending appropriate countermeasures. The decision to implement those recommendations and mitigate the risk, or to accept risk, is that of the security management in consultation with organizational leadership. The security manager, in consultation with organizational leadership, is responsible for identifying and implementing the most cost-effective countermeasure appropriate for mitigating a vulnerability, thereby reducing the risk to an acceptable level.

The risk assessment can be processed in a qualitative or quantitative framework. The qualitative risk assessment will be based on methods, principles, or rules for assessing risk based on nonnumerical categories or levels. An example would be where the security manager assigns categorization of risks as low, medium, or high based on known data or assumptions regarding the risk environment.

A quantitative risk assessment will utilize a set of methods, principles, or rules for assessing risks based on the use of numbers where the meanings and proportionality of values are maintained inside and outside the context of the assessment. For example, a security manager could use a quantitative risk assessment methodology to assess the risk of loss based on statistical past loss data. While a semiquantitative methodology also involves the use of numbers, only a purely quantitative methodology uses numbers in a way that allows for the consistent use of values outside the context of the assessment. To make an informed risk-based decision regarding the mitigation or the acceptance of risk, collaboration between the security manager and organizational leadership is required. For any countermeasure that is recommended, the security manager must provide all information pertinent to the decision, such as the nature of the threat, the specific vulnerabilities, an understanding of the potential consequences, and the costs.

ASSESSING RISK

The security manager and organizational leadership need accurate information in order to make effective decisions. Security management must have the authority, appropriate clearance, and access to expert resources to gain an understanding of risk to render sound decisions. This requires something beyond just an understanding of the security issues. It also requires a knowledge of the mission and priorities of organizational leadership and the associated cost implications. This approach promotes comparability and a shared understanding of information and assessment in the decision process, and facilitates better structured and informed decision making. The risk assessment is a methodical process and it should:

1. Define the organizational asset and its context
2. Identify risks to the organizational asset
3. Determine asset vulnerability to identified risks
4. Develop multiple approaches to deal with organizational risk
5. Decide and implement the optimal approach to deal with asset risk
6. Monitor and evaluate the risk mitigation measure(s)
7. Communication (DHS, 2011a)

DEFINE THE ORGANIZATION AND ITS CONTEXT

To execute a risk assessment, it is critical to define the context that the risk management effort will support. The security manager must gain an understanding of the environment in which the risks are to be managed, taking into account all organizational concerns and risk tolerance. Defining the context will inform and shape successive stages of the risk management cycle.

To determine the necessary level of protection to adequately mitigate risk, a security manager must understand the role assets play in an organization. Variations in the nature of mission, location, and physical configuration of an organization may create unique risks or risks that are relatively higher or lower in some cases than with other organizations. The baseline level of protection may not address these risks appropriately. It may provide too little protection leaving an unmitigated risk. Or, it may provide more protection than is necessary resulting in the expenditure of resources where they are not needed. This might reduce the availability of resources that could be applied elsewhere.

This initial step of clearly defining goals and objectives is essential to identifying, assessing, and managing the areas that may impact the success of your risk assessment. This concerns organizational leadership, the mission, and values of the organization.

This also should take into consideration existing risk management tools, including organizational policies and procedures. The security manager must understand the decision-making process in the organization and the commitment of organizational leadership to support the risk assessment process.

Organizational leadership and staff must be invested in the risk assessment to assure their support and commitment to the process. The security manager also should be sure that organizational leadership grasps the technical aspect of the process so they will not only understand, but also buy into the risk assessment process. This initial portion of the risk assessment process requires communication between the risk manager and all parties involved in the process.

In this initial step of risk assessment, the security manager must identify the staff (including their skill levels), resources (financial), and other organizational resources available for risk assessment efforts. The security manager must be flexible and execute a great deal of understanding as he/she moves into the risk assessment process. The inability of the security manager to recognize aspects of organizational culture and decision makers can have a negative impact on the success of the process.

The security manager also must consider existing security practices and the overall culture of security in the organization. He/she should define in detail the following key areas of security management:

- Current asset classification practices
- Existing assignment of roles for organizational ownership of assets
- Existing asset responsibilities, including tasks and procedures, to be followed by the staff members who work with the asset
- Existing policies regarding the securing of organizational assets, including how violations are reported and dealt with
- Existing security awareness practices, including staff educational and training programs related to asset security

IDENTIFY RISKS TO THE ORGANIZATIONAL ASSET

Identifying a preliminary list of risks can generally be done from a basic knowledge of the asset, its function in the organization, and an understanding of its vulnerabilities. This can be completed by understanding the asset, its function relative to the organization's goals and objectives, then determining the risks, hazards, resources, and organizational vulnerabilities that impact them, which could result in risk. This method will provide a list of potential loss situations to the organizational asset.

Risks to organizational assets come from both internal and external vulnerabilities. Examples of internal risks often surround actions (or inaction) by staff and failures in organizational systems. All organizations are vulnerable to internal risks. This step of the risk assessment will identify weaknesses in the organization's internal systems and processes.

External sources of risk can be caused by many factors. External factors resulting in vulnerability can run a gamut: global, political, and societal trends, as well as hazards from natural disasters, terrorism, cyberspace, pandemics, transnational crime, and manmade accidents. It is the role of the security manager to think critically and recognize all the potential for loss.

DETERMINE ASSET VULNERABILITY TO IDENTIFIED RISKS

The purpose of this step is to utilize a methodical framework to objectively determine asset vulnerability to identified risks, which involves:

- Determining a formal risk assessment methodology
- Gathering data
- Executing the risk assessment methodology
- Validating and verifying the data
- Analyzing the outputs

Determining a Formal Risk Assessment Methodology

In determining a formal risk assessment methodology, the security manager should try to keep it as simple as possible while providing as comprehensive and accurate data as possible. The more simplistic any process, the less likely it is for failure or misinterpretation. Simplicity and practicality will have an inherit value when you are "selling" your security program to organizational stakeholders. This will allow you to better explain technical loss data to both laymen and experts.

Risk assessment methodologies can be either qualitative or quantitative, but, when well-designed, both types of assessments can provide results for a valid understanding of the asset-risk environment. Failure to choose a valid risk assessment can result in the process being flawed, if not completely unreliable.

Gathering Data

The security manager should gather enough data to provide practical and valid data in the assessment. There are a number of potential sources for risk information. Some of the most commonly used include staff interviews, historical data, models, simulations, and subject matter experts.

Many pieces of data are not known precisely. When an organization has not experienced a particular negative event, it may be oblivious to that risk and loss potential. It is crucial that the security manager follow critical thinking guidelines. The assumptions and uncertainty in the inputs should be considered in each step of the assessment's methodology to determine how they affect the outputs. Uncertainty in the outputs should then be communicated to the decision maker as well as the assumptions that underpin the analysis. It also is useful to consider the impact of the uncertainty and how sensitive the assessment of risk is to particular pieces of uncertain data.

Executing the Risk Assessment Methodology

The risk assessment methodology the security manager chooses to use is the framework that guides how the data are processed to provide valid results for decision making. There are many different methodical frameworks that can be used to provide the security manager with an understanding of the risk environment and subsequent methods of protection and control to enhance security. In the process of implementing the risk assessment, the security manager should critically evaluate the assessment tool to assure its validity in providing the information needed.

As the assessment process is implemented, the security manager should begin to gain a justifiable understanding of the risk environment. It is within this process that the security should develop and evaluate alternative means of protection as well. While approaches for developing and evaluating alternatives are as diverse as the problem sets, considerations may include:

- Reviewing lessons learned from relevant past incidents
- Consulting subject matter experts, best practices, and governmental guidelines
- Brainstorming with stakeholders
- Organizing risk management actions
- Evaluating options for risk reduction and residual risk

Validating and Verifying the Data

The security manager should continually validate and verify the data collected to assure its soundness. The importance of validating and verifying data is fundamental to the risk assessment process because it is the foundation for making decisions to protect an asset, among a number of alternatives in an uncertain environment. The key moment in the execution of any risk management process is when a decision maker chooses among alternatives for managing risks and makes the decision to implement the selected course of action. This can include making an affirmative decision to implement a new alternative as well as the decision to maintain the status quo.

Analyzing the Outputs

In analyzing the outputs from the risk assessment, the risk manager should evaluate and monitor performance to determine whether the implemented risk management options can achieve the goals and objectives of the organization. In addition to assessing performance, organizations should guard against unintended adverse impacts, such as creating additional risk or failing to recognize changes in risk characteristics.

The evaluation phase is designed to bring a systematic, disciplined approach to assessing and improving the effectiveness of risk management program implementation. It is not just the implementation that needs to be evaluated and improved, it is the actual risk reduction measures themselves. Evaluation should be conducted in a way that is commensurate with both the level of risk and the scope of the mission.

In practice, the determination of risks to assets rarely occurs linearly. Instead, security managers often move back and forth between the tasks, such as refining a methodology after some data have been gathered.

DEVELOP MULTIPLE APPROACHES TO DEAL WITH ORGANIZATIONAL RISK

The security manager should develop multiple approaches to deal with organizational risk. The analysis of the risk assessment process is often enhanced by developing various models and techniques from various loss scenarios. The utilization of differing approaches is based on assumptions regarding how potential risks materialize against organizational assets.

The risk assessment process should yield multiple process and protection measures for enabling timely and relevant mitigation of risks by monitoring predictive indicators, escalating information on increased risk exposures, and making risk-informed decisions in an integrated manner. The security manager also should analyze the impact of these risk mitigation measures while considering the costs and benefits of the alternatives.

DECIDE AND IMPLEMENT THE OPTIMAL APPROACH TO DEAL WITH ASSET RISK

Risk management entails making decisions about the best options among a number of alternatives in an uncertain environment. The key moment in the execution of any risk management process is when the security manager, in cooperation with organizational leadership, chooses among alternatives for managing risks, and makes the decision to implement the selected countermeasure(s). This can include making an affirmative decision to implement a new alternative as well as the decision to maintain existing security measures.

Cost considerations can be a key factor in decisions to implement countermeasures. Cost-benefit analysis is a cost analysis methodology used to justify security expenditure. The cost-benefit analysis is a comparative assessment of the benefits from your security measure and the costs to perform it, in relation to the financial impact resulting from potential loss to the asset. The security manager should consider costs for all phases of the security project. They may be one-time capital costs or recurring costs. The determining criterion in a cost-benefit analysis is that the security benefit exceeds the cost. The amount that benefit exceeds the cost is based on the security manager in consultation with organizational leadership.

MONITOR AND EVALUATE THE RISK MITIGATION MEASURE(S)

The security manager should be monitoring implemented security measures and comparing effects to help influence subsequent risk management alternatives and decisions. The risk assessment will identify possible risks and the likelihood of occurrences rated in terms of impact or severity and probability. This enables the security manager to have the data to develop response strategies and allocate resources appropriately. Security management then ensures that risk assessments become an ongoing process, in which objectives, risks, risk mitigation measures, and controls are regularly reevaluated.

The security manager should evaluate and monitor performance security measures to determine whether it is achieving its goals and objectives. In addition to assessing performance, the security manager should monitor for unintended adverse impacts, such as creating additional risk or failing to recognize changes in risk characteristics as a result of implementation of the security measure(s).

The evaluation phase is designed to bring a systematic, disciplined approach to assessing and improving the effectiveness of risk management implementation. It is not just the implementation that needs to be evaluated, it is the actual risk reduction measures themselves. Evaluation should be conducted in a way that is commensurate with both the level of risk and the scope of the organizational mission.

RISK COMMUNICATIONS

The foundation for each element of the risk management process is effective communications with organizational leadership, staff, other stakeholders, and customers. Consistent, two-way communication throughout the process helps ensure that all of those involved share a common understanding of asset risk and factors involved in managing it. Effective communication is an essential element in executing plans and countermeasures and in explaining risks and security management decisions. Such external communications should occur throughout the security management process and should be considered integral to effective risk management.

CONCLUSION

The primary role of a security manager is to organize the countermeasures that are generally considered to mitigate the risk from a particular undesirable event. Security managers are aided by risk assessments to help them in understanding organizational risk. A risk assessment identifies a generic set of undesirable events that may impact organizational assets, and risk management relates them to the applicable security measures.

By determining the necessary level of protection according to a risk assessment, it is possible to ensure that the most cost-effective security program is implemented without waste or lingering vulnerability. Further incorporating a cost-benefit assessment will provide additional insight for the security manager. The security manager should determine whether the countermeasures adequately mitigate risk in an economically acceptable manner.

In all cases, the risk management process should document clearly the reason why the security measure(s) are necessary. It is extremely important that the rationale for accepting risk be well documented, including alternate strategies that are considered or implemented, and opportunities in the future to implement the necessary level of protection, with these findings communicated to all stakeholders in the process.

EXERCISES

The risk assessment is a methodical process that promotes comparability and a shared understanding of information and assessment in the decision process, and facilitates better structured and informed decision making. What are the seven steps of the formal risk management process and how would you apply them to a local grocery store. Give examples for each of the steps.

REFERENCES

U.S. Department of Homeland Security. 2008. DHS risk lexicon. Online at: http://www.dhs. gov/xlibrary/assets/dhs_risk_lexicon.pdf
U.S. Department of Homeland Security. 2011a. Applying risk management principles to guide federal investments. Report to the Senate Committee on Homeland Security and Governmental Affairs. U.S. Government Accountability Office. Online at: http://www. gao.gov/new.items/d07386t.pdf
U.S. Department of Homeland Security. 2011b. National preparedness: DHS and HHS can further strengthen coordination for chemical, biological, radiological, and nuclear risk assessments. Report to the Senate Committee on Homeland Security and Governmental Affairs. U.S. Government Accountability Office. Online at: http://www.gao.gov/new.items/d11606.pdf.

10 Computer and Information Security

Many organizations collect large amounts of information. The information can range from employee records, operational data, customer databases, trade secrets, etc. Organizational information commonly collected may contain administrative reports documenting various types of data for day-to-day operations or other supplementary support. The fact is that most organizations retain a lot of information in their operational processes.

Information is a vital resource for an organizational staff to utilize in planning programs and services, and for scheduling and completing reports for organizational stakeholders, if not for local, state, and federal agencies. In emergency situations, information must be readily available to make a valid response to a life safety event. Organizations must maintain information integrity, accessibility, and confidentially.

Organizational information is a compilation of records, files, documents, and other materials that may contain many sensitive pieces of data. Information may be kept in a variety of formats including handwritten, printed, digital files, and video or audio recordings. Information must be accurate and available to make timely decisions. In emergency situations, accurate and accessible information may play an important role in safety. Furthermore, organizations' obligations could have a legal consequence if they fail to safely maintain information.

This chapter has three primary purposes. First, it is to serve as a guide to help organizations look at information and give it a value. A risk assessment is a methodical tool to help practitioners look at information as assets to the organization and determine a value of that artifact and, then, understand the threats and vulnerability as the basis for protection. Second, we will address legislative requirements outlining methods for keeping information safe. Finally, fundamental principles in applying policy and procedures in the pursuit of safe keeping of information will be addressed.

SAFEGUARDING INFORMATION

Protecting information requires maintaining confidentiality and integrity, while ensuring accessibility and availability of information via security, safety, and ethical behavior. Confidentiality involves the prevention of disclosure of information to unauthorized individuals. Information integrity means that data cannot be modified without someone being able to detect its modification. Information availability suggests that it must be accessible when it is needed. The information is made safe by providing physical and procedural measures, ensuring that only those with specific needs have access to the information, and by promoting an ethical understanding to all involved as to the sensitivity of information.

Organizations should develop plans to safeguard information. Developing a comprehensive information safety plan requires going through a process to assess information value, vulnerability, and risk. This process is called a risk assessment or risk analysis. The risk assessment is a continual and dynamic process because risk management is an ongoing iterative process. Risk assessment must be repeated indefinitely. Any time new technology is introduced or changes are made to informational processes, the organization's information environment changes. These changes present the possibility of new threats and vulnerabilities impacting the safekeeping of information. Therefore, the risk must be reevaluated. The organization must constantly balance countermeasures and controls to manage risk to information. Information safety involves balancing productivity, cost effectiveness of the countermeasure, and the confidentiality, integrity, and availability of the information.

An information risk assessment is a variation of a more traditional risk assessment, which would focus on identifying assets and looking at the threats and vulnerabilities of each asset to potential loss. In conducting an information risk assessment, the first step is to identify information assets and estimate their value to the organization.

Identifying information assets requires investigation and asking questions, such as: What type of information does the organization maintain? What is its value to the organization? Where is the information stored? Is it stored digitally or in hard copies?

Information includes all of the data maintained by organizations and by other parties acting for the organization. Information collected by an organization about staff often includes personal information, such as pay data, Social Security number, pictures, or a listing of personal data that would have varying degrees of sensitivity, such as records pertaining to medical and health information. This also should be addressed as sensitive information. The organization will maintain information as well that may be of a more sensitive nature. This could include information specifically related to organizational processes.

The security manager should identify information, and also place a value on it. The security manager should calculate the value of each information asset. This would require a qualitative or quantitative analysis of information to determine what that information is worth to the organization. Questions that should be considered include: How can we meet the primary mission of this organizational enterprise in the absence of this information? What legal ramifications exist if we do not safely maintain this information artifact?

The second step is to conduct a threat assessment to the organization's information. A threat assessment of information is a daunting task for educators. It requires the knowledge of all the information maintained by the organization and potential threats to that information. Threats can be obvious, such as natural disasters or, more ambiguous, such as a computer virus. Information stored in different mediums may have completely different characteristics and subsequent threats. The threats to the file cabinet in the secretary's office are much different from organizational information stored on the network cloud.

A threat assessment should look at each information asset and then determine its liability. Threats can include both accidental and malicious acts originating from inside or outside the organization. The threat assessment dictates that each situation of concern be viewed and assessed individually. Application of the assessment is guided by the facts of the specific threat and carried out through an analysis of its characteristics. Providing a safe information environment is very different and is based on where the information is stored.

Threats to digital information are viruses, malware, exploited vulnerabilities, remote access, mobile devices, social networks, and cloud computing. Viruses and malware can render information useless. Most active computer users have fallen prey to viruses and malware that damage and destroy information. Viruses are malicious codes that cause an infected computer to spread the virus to other associated computers via the network or email contact lists. Malware, while maybe destructive or disruptive in nature, will not have the inherent ability to spread itself. So, while you may get a malware infection by visiting a Web site that hosts the malware, it is not actually a virus unless it utilizes your contact list or network directory services to propagate itself. Either the malware- or virus-infected machine or network can make an organization's information useless (Data Security, n.d.)

A vulnerability exploit is another threat to an organization's digital information. A vulnerability exploit is where a "hole" is found in applications, or systems software, that facilitates unauthorized access to information. It will allow an unauthorized user to access information and data that he/she does not have the legitimate right to access. Vulnerability exploits are often utilized by viruses and malware to penetrate networked computer systems.

Mobile devices are another threat to information safety. Storing and removing sensitive information from an organization on a USB drive and then losing the drive poses a threat. Another good example is the laptop computer. Organization-owned laptops often house sensitive information. The loss of that information to theft is a major cause of the loss of information integrity. When computer information is outside the organization's physical network, its threat changes. Having an organization-owned laptop (with organizational information embedded) outside the logical and physical security of the business poses additional threats from both loss and exploitation, and must be considered in a threat assessment.

Online social networking poses additional threats to information safety. Social networking sites, such as Facebook and Twitter, can pose serious threats to organizational information both directly and indirectly. Social networking sites are breeding grounds for viruses and malware plus other attacks to organizational information systems. Social networking sites also can allow organizational staff and stakeholders to post sensitive information.

Organizations using cloud-based resources must be aware of threats caused by utilizing that resource. The cloud is an Internet-based network that allows anyone to access applications on a network platform that is leased or rented to the organization similar to a utility or tenement arrangement. The cloud extends the system so staff and stakeholders have 24/7 access to not only their email, but also personal files and applications software anywhere they have a computer and Internet access. For example, digitized information does not have to be installed on the remote computer nor does it have to be connected to a local area network to function. The application is run from the cloud and data generated is stored on the cloud, which can be accessed anytime by anyone with Internet service and proper access credentials.

Staff may unwittingly pose a major threat to organizational information. This could range from a desperate and disgruntled employee who was just fired because of performance to the careless employee who unknowingly releases sensitive information. There is no way to completely eliminate the threat of legitimate insiders, but, through good safety policies and procedures, information loss could be greatly minimized. Careless and untrained employees could unknowingly release sensitive, if not damaging information about an organization. Policies, procedures, training, and technical measures can play a major role in reducing an organization's information threats (McCallister, Grance, and Scarfone, 2010)

The threat assessment of information is a huge task. It requires the knowledge and skills to address threats to the information maintained by organizations. Regardless if a threat is from natural disasters or a computer virus, it must be addressed. And, once the threat is addressed, the organization should evaluate the vulnerability of information to that loss.

The third step is to conduct a vulnerability assessment to organizational information. A vulnerability assessment should calculate the probability that a threat to information assets could occur. The vulnerability assessment could be based on quantitative or qualitative evaluation of threats to information. The resulting data should give you an idea of the likelihood that loss of confidentiality, integrity, or access may occur so that measures of information safety could be implemented (Data Security, n.d.)

Often the creation of a chart or matrix (Table 10.1) will facilitate professionals in determining threats and vulnerabilities to information. Either qualitatively or quantitatively staff can value information and the threat vulnerability (Elky, 2006).

For example, for each informational asset, give it a value from low to high. Then, for each threat vulnerability to the information, give it a value from low to high. The intercept of these characteristics should provide the data to make a decision regarding how the informational asset should be protected from that threat.

An overly simplistic example is the threat of electrical surges to render information stored on a computer irretrievable. In some parts of the country, this could be a very high threat vulnerability and the information could be very highly valued. In these situations, it would be worth spending $20 for a surge protector. However, is the value justifiable in buying a $500 surge protection system? From the data in this

TABLE 10.1

Determining Threats and Vulnerabilities to Information

High Threat Vulnerability			High Risk
Medium Threat Vulnerability		Medium Risk	
Low Threat Vulnerability	Low Risk		
	Low Information Value	Medium Information Value	High Information Value

scenario, we do not know if we can justify spending more money because this will have to be balanced in terms of a comprehensive information safety program and organization budgets. The point is that, with the risk assessment, we can make value judgments based on more than assumptions.

The data from an accurate risk assessment of information value, threats, and vulnerabilities will provide the security manager the ability to identify, select, and implement appropriate security measures. This does not require that the security manager has to be an expert in safeguarding information, but it does assume he/she has knowledge of information and knows when to provide a proportional response, even if that requires getting expert assistance in completing the task. The security manager should understand productivity, cost effectiveness, and value of the informational asset and its threats and vulnerabilities and, subsequently, how to go about protecting it (Elky, 2006).

Implementing a plan to keep information safe in an organization is a balancing act. Security managers must evaluate the effectiveness of measures ensuring security without discernible loss of productivity. Providing information security is contingent upon the security manager understanding the environment, risks, and vulnerabilities of technology. Because of the breadth of the issues and complexity of the environment, there is no more valid approach to achieving an information-safe environment in organizations than to promote secure, safe, and ethically sound informational practices by all staff and stakeholders. An organization must have both policies and procedures to provide for authentication, firewalls, and virus protection on its computer systems.

Authentication will provide secure login and allow individuals to access only the data and applications they require. The firewall also serves as an access control device that will limit the user access to unwanted programs and data. Virus protection programs will protect users from viruses and malware on the network. Each of these three areas must be addressed by computer professionals in that environment. The best role for the security manager is to be aware of the value of information and its associated risks to develop policies and procedures to protect it. For the security manager, this means working with a service provider or consultant because they usually have access to greater levels of expertise (Elky, 2006).

Information security in organizations is extremely necessary because of the liabilities of improperly handling information. In many organizations, federal regulations require that they must make information safe regarding who will interact with it and how. Organizations must understand areas of compliance and how to safeguard the standards and regulation requirements.

FAMILY EDUCATION RIGHTS AND PRIVACY ACT (FERPA)

Schools, for example, have comprehensive information security measures mandated by the Family Education Rights and Privacy Act of 1974 (FERPA), which is a federal law that protects the privacy of student education information. Students have specific, protected rights regarding the release of information, and FERPA requires that institutions adhere strictly to its guidelines. Therefore, it is imperative that the faculty and staff have a working knowledge of FERPA guidelines before accessing organization information.

With FERPA, educational information is put into two broad categories: directory information and nondirectory information. Each category of educational record is afforded different safety protections. Therefore, it is important for faculty and staff to know the type of educational record that is being considered for disclosure (National Association of Colleges and Employers, 2008).

Directory information is the student's educational record and is generally not considered harmful or an invasion of privacy if released. Under FERPA, the organization may disclose this type of information without the written consent of the student. However, the student (or guardian, until a student is 18 years of age or attends a postsecondary institution) can restrict the release of directory information by submitting a formal request to the school. Directory information includes (National Association of Colleges and Employers, 2008):

- Name
- Address
- Phone number and email address
- Dates of attendance
- Degrees awarded
- Enrollment status
- Major field of study

Educational organizations should always make students and parents aware that such information is considered by the organization to be directory information and, as such, may be disclosed to a third party. Organizations, however, should make it very clear to students and parents that they can prevent the release of directory information (McCallister, Grance, and Scarfone, 2010) by request.

Nondirectory information is any educational record not considered directory information. Nondirectory information must not be released to anyone, without consent of the student, parents, or guardian until a student is 18 years of age or attends a postsecondary institution, and then at the consent of the student. FERPA insists that faculty and staff can access nondirectory information only if they have a legitimate academic need to do so. Nondirectory information includes (Federal Register, 2001):

- Social security number
- Student identification number
- Race, ethnicity, and/or nationality
- Gender
- Transcripts
- Grades

Student's distinct identifiers, such as race, gender, ethnicity, grades, and transcripts are nondirectory information and, therefore, are protected educational records under FERPA. Students have a right to privacy regarding information held by the school. Organizations must ensure that students have privacy of this information under FERPA. FERPA also gives a student and parents (if the student is under 18) the right to access educational information kept by the school. It also gives the right to limit educational information being disclosed, the right to amend educational information, and the right to file complaints against the organization for disclosing educational information in violation of FERPA (Federal Register, 2001).

Students and parents have a right to know about the purpose, content, and location of information kept as a part of their educational records. They also have a right to expect that information in their educational records will be kept confidential unless they give permission to the organization to disclose such information. Therefore, it is important to understand how educational information is defined under FERPA (National Association of Colleges and Employers, 2008).

Prior written consent is always required before institutions can legitimately disclose nondirectory student information. In many cases, organizations tailor a consent form to meet their unique educational needs. However, prior written consent must include the following elements (Federal Register, 2001):

- Specify the information to be disclosed
- State the purpose of the disclosure
- Identify the party or class of parties to whom the disclosure is to be made
- The date
- The signature of the student, parent, guardian whose record is to be disclosed
- The signature of the custodian of the educational record

Prior written consent is not required when disclosure is made directly to the student or to other organization officials within the same institution where there is a legitimate educational interest. A legitimate educational interest may include enrollment or transfer matters, financial aid issues, or information requested by regional accrediting organizations (National Association of Colleges and Employers, 2008).

Institutions do not need prior written consent to disclose nondirectory information where the health and safety of the student is at issue, when complying with a judicial order or subpoena, or where, as a result of a crime of violence, a disciplinary hearing was conducted by the school, a final decision was recorded, and the alleged victim seeks disclosure. In order for institutions to be able to disseminate nondirectory information in these instances, FERPA requires that institutions annually publish the policies and procedures that the institutions will follow in order to meet FERPA guidelines (McCallister, Grance, and Scarfone, 2010).

The Family Education Rights and Privacy Act was enacted by Congress to protect the privacy of student educational information. This privacy right is a right vested in the student. Institutions may disclose directory information in the student's educational record without the student's consent. Institutions must have written permission from the student in order to release any information from a student's educational record (McCallister, Grance, and Scarfone, 2010).

OTHER REGULATORY ACTS REGARDING INFORMATION

Information is protected by other agencies and groups in addition to FERPA. The Departments of Agriculture, Health and Human Services, and Justice defend information privacy and apply it to the information in organizations. State and local entities may require safeguards for handling information in a safe and secure manner. Professional standards of ethical practice of organization, such as doctors and nurses, psychologists, and other professionals, operate, may also establish privacy restrictions (McCallister, Grance, and Scarfone, 2010).

Drug and alcohol prevention and treatment services for are covered by confidentiality restrictions administered by the U.S. Department of Health and Human Services. Some states have regulations regarding employee rights to seek treatment for certain health and mental health conditions, including sexually transmitted diseases, HIV testing and treatment, pregnancy, and mental health counseling. Regulations also protect information pertaining to HIV confidentiality, medical information, child abuse, privileged communications, and state-specific information retention and destruction regulations (Privacy Rights Clearinghouse, 2012).

MAKING INFORMATION SAFE

Information can be a compilation of records, files, documents, and other materials that contain information maintained by organizations. Likewise, information is a vital resource for organizations in planning, operations, and in providing services, and organizations can be legally and ethically challenged to maintain information integrity, accessibility, and confidentiality. Information may be kept in a variety of formats including handwritten, printed, digital files, and video or audio recordings. Organizations have an obligation to maintain information integrity, accessibility, and confidentiality, and information must be accurate and available to make timely decisions. The organization's obligation could have legal consequence if it fails to safely maintain information of students (Privacy Rights Clearinghouse, 2012).

Organizations have an ethical responsibility to keep personal information about its students safe. Public organizations are further bound by federal legislation that provides mandates regarding information. Therefore, it is imperative that organizations have a plan regarding information security.

The final part of this chapter addresses the role of the organization in providing a secure information environment. Providing information security is contingent upon security managers understanding the environment, risks, and vulnerabilities of technology. This chapter also assumes that most organizational information is digital. Because of the breadth of the issues and complexity of the environment, there is no more valid approach to achieving a safe information environment in organizations than to use a multimode approach, which will promote secure, safe, and ethically sound practices. Organizations can make information secure by providing security, safety, and awareness of anyone who deals with this information. The most common approach is to use layers to protect information. There are specific tools that can be utilized at each of these layers.

Instituting security will require specific technical information that most organizational stakeholders will not possess. Because of this, most organizations will need an expert to install and configure these applications and appliances. Security layers include firewalls and routers. A firewall will stop unwanted applications and data from entering an organization's network. Unwanted applications and data could be anything from known viruses to accessing social networking sites. The firewall can block all network traffic that is unwanted. If you do not want Facebook to be accessed from organizational computing resources, then add it to the blocked list.

The routers must be protected on a network. They are like a train station through which all information passes. Routers direct information on where it needs to go. The router is like a firewall because you can stop information from being dispersed into the network; however, its primary function is to direct network traffic both in and out of the local network. From this perspective, a firewall or a router can both stop unwanted applications and data, but only the routers and switches show data where it needs to go on the network. The router is a very critical link for a network.

Network controls are a third primary layer in addressing information safety. Network controls include authentication and file sharing controls. Authentication will provide a user name and password authentication system. This means that only authorized users have access to the network and its information. The file sharing controls set an access level for each authenticated user to access applications and information on the network. From this perspective, you only give users the rights to information that is necessary for their task. In respect to information, you can give users rights to read, write, and modify discriminately for each information item stored on the network. The keys for network controls are to have a good password system and only give users the rights to what they need.

Software layers of safety will include virus and malware detection/prevention systems. Malware and viruses can be one of the biggest problems for digitally stored information. The best way to prevent losses from these threats is by maintaining adequate antivirus and malware protection software and keeping all updates installed on the organization's computers. Antivirus and malware software must be updated frequently. Often a new virus cannot be stopped until the antivirus software has the correct definition to stop or quarantine it. Often times, an update to a computer's operating system and applications software is required because the manufacture has found vulnerability in the system that must be addressed through updates and patches.

The final layer of information security involves creating informed information custodians. Information users should understand the value of information to the organization. Through training and continual reinforcement, information will be respected and cared for to assure its integrity, accessibility, and confidentiality. Everyone using organizational information must adhere to an acceptable use policy with consequences for violating it. Instill a sense of information responsibility. The best time to begin establishing a culture of information security is now.

It is not the intent of this section to address information security from the view of a network administrator, but to make the security manager aware of what is needed to create and maintain an information-safe environment in the organization. Organizational stakeholders should understand a fundamental process, such as an information risk assessment, to view value, threats, and vulnerability of information. An organization should be aware of all federal, state, and local laws regarding the use and maintenance of information. And, finally, organizations must have both policies and procedures in order to institute safety systems for authentication, firewalls, and virus protection of computer systems. Only through these measures can organizations adequately address information security.

EXERCISES

Sensitive personal information of 105 Bolder Machine Works employees, including names and Social Security numbers, were accidentally emailed to several customers Monday. Bolder Machine Works director's office released the information of the employees via an email attachment sent to 26 recipients. Plant manager Mitch Johnson informed the employees whose information was disclosed: "We take the safety of employee information very seriously, and we deeply regret this error," Johnson said. "We are reviewing the incident and are taking steps to prevent similar incidents from occurring in the future. This should not have happened, and we apologize for it."

On further investigation, the manager determined the information was sent to a group of customers as an attachment when the intended file attachment was accidentally replaced with the employee data file by an administrative assistant. It was an honest mistake.

- As a security manager, how would you address this issue?
- What factors must be considered in addressing the issue?
- What can be done to make sure this does not happen again?
  1. What data will a risk assessment of information provide?
  2. What is the layered approach to information security and identify components or layers?
  3. Without federal and state statutes requiring compliance, how do you think information security should be addressed in organizations?

REFERENCES

Data Security Checklist. n.d. Online at: http://nces.ed.gov/programs/ptac/pdf/ptac-data-security-checklist.pdf (accessed November 15, 2011).

Elky, S. 2006. An introduction to information system risk management. Online at: http://www.sans.org/reading_room/whitepapers/auditing/introduction-information-system-risk-management_1204 (accessed November 15, 2011).

Federal Register. 2001. 34 CFR Part 99, Part V, Family Education Rights and Privacy, Final Rule. Office of Family Policy Compliance, Family Education Rights and Privacy Act (FERPA), July 26.

McCallister, E., T. Grance, and K. Scarfone. 2010. Guide to protecting the confidentiality of personally identifiable information (PII), NIST Special Publication 800-122. Online at: http://csrc.nist.gov/publications/nistpubs/800-122/sp800-122.pdf (accessed November 15, 2011).

National Association of Colleges and Employers. 2008. FERPA primer: The basics and beyond. Online at: http://www.naceweb.org/public/ferpa0808.htm (accessed November 15, 2011).

Privacy Rights Clearinghouse. 2012. Fact sheet 29: Privacy in education: Guide for parents and adult-age students. Online at: https://www.privacyrights.org/fs/fs29-education. htm#11 (accessed November 15, 2011).

11 Cybersecurity

Cybersecurity is a topic that has spawned from the computer security domain since the profusion of Internet-based systems. Businesses rely on a vast array of computer networks to communicate, plan, provide services, and, literally, run our economy. The *cyber* aspect denotes the ability to remotely, if not virtually, access an organizations networked computer systems.

Cyber intrusions and attacks on networked computer systems have increased dramatically over the past decade, exposing sensitive personal and business information, disrupting critical operations, and imposing high costs on the economy (Department of Homeland Security, n.d.).

There is an increased vulnerability to network systems that have been classified as critical infrastructures, such as financial systems, chemical plants, and water and electric utilities. The role of security management in dealing with these environments often becomes a shared responsibility through various organizational units. Considering cybersecurity takes a multidimensional approach because most organizations have administrative technology systems to retain everything from employee data to inventories to process data used in the daily operation of the business. And, challengingly, a security manager will have a role in facilitating a secure cyber environment because, logistically, cybersecurity extends from the organization's own local area network to the far-reaching network of the Internet.

Computers and cyber technologies enhance organizational performance by engaging, involving, and empowering stakeholders. Further diffusion of Web 2.0 technologies and smartphones provide another interesting dimension to cybersecurity. Staff, as well as organizational leadership, have the ability to access, communicate, and process data via a robust network platform that they can hold in their hand or store in their pocket. Web 2.0 has made the Internet an instantaneous–participatory–interactive–community differing from the initial incarnation of the Web as an environment where you could gather (or post) information and communicate via asynchronous systems. Smartphones, utilizing mobile computer operating systems, provide the user the ability to interact via Web 2.0 applications synchronously. Web 2.0, combined with smartphones, has provided advanced computing and communication tools to its users and an additional challenge to maintain cyber-secure environments.

Organizations are increasingly being challenged to maintain a technological environment that is cyber secure for all stakeholders as well as for staff and leadership. In addressing the assurance of a cyber-secure environment, the first section of this chapter will be viewed from three cyber environments: the Internet, social networking, and smartphones, each with their own purpose, attributes, significance, value, and subsequent risks. The second section focuses on creating a cyber-secure environment by implementing measures to cultivate secure and ethical cyber practices.

Today's developing information-age technology has intensified the importance of cybersecurity as critical as physical security in protecting many organizations. The Energy Sector (Electric, Oil, and Gas) has rapidly responded to the increasing need for enterprise-level cybersecurity efforts and business continuity plans. Vulnerability assessments have not only improved security, but also have demonstrated organizational commitment to secure organizational environments through cybersecurity.

It is not the intent of this chapter to provide readers with the ability to develop a cyber-secure environment in organizations. Technical aspects, such as firewalls, malicious code, and access controls, need to be addressed by network administrators who can configure these protection systems properly. However, it is proposed that this chapter enlighten security managers with the knowledge base that will allow them to understand the various factors that construct the organization's cyber environment, so they can provide an understanding to the approach to reducing an organization's cyber risks. Minimizing cyber risks in organizations can be best addressed by making staff and stakeholders aware of their role in the process. Because of organizations' and educators' place in the lives of students, it is imperative that cybersecurity be a portion of the organizational policy process.

THE INTERNET

The Internet is a worldwide public computer network and was originally founded by the Department of Defense in the 1970s using Transmission Control Protocol/Internet Protocol (TCP/IP) to connect computers and networks. Since the inception of Hyper Text Markup Language (HTML) in the early 1990s, the Internet evolved from a text-based communications platform to graphically interfaced Web pages. HTML fueled the development of the World Wide Web (WWW) into Web sites that would be publicly accessible, hosted via Internet-connected network servers, and would allow for other media files rather than mere text to be accessed.

The use of the Internet in organizations grew rapidly after the inception of HTML. The Internet has become a paradox to the organization, which is that, while the Internet is an environment that is beneficial to modern business systems, it also contains many risks and hazards to organizations. The Internet is a great resource, full of endless amounts of information and resources. Staff use of email has diffused throughout all organizations in the past 20 years. Staff Internet access is not only in the organization setting, research shows 82% have broadband Internet access at home (Rainie, 2011). While the Internet provides many positive attributes for organizations, it is easy to find sites that could have a negative impact.

EMAIL

Email in the educational environment is frequently used by staff and other stakeholders for sending and receiving electronic messages. Email allows employees to keep in touch with family, friends, peers, as well as for business-related communication.

Potential risks with email relate to the inherent quality of the Internet. It utilizes a public network architecture where you can communicate with other email users indiscriminately. There is no required validation that users are who they say they are.

Furthermore, email users can communicate with others through unsolicited messages. These unsolicited messages (or spam) can vary from sexually explicit material to products for sale to moneymaking schemes or to a host for a malicious program.

Another facet of email involves a substantial movement by organizational systems to use cloud-based network systems. Cloud-based services allow users to access applications on a network platform that is leased or rented to organizations, similar to a utility or tenement arrangement. The cloud extends the system so staff have 24/7 access to not only their email, but also shared files and applications software anywhere they have a computer and Internet access. For example, some cloud-based applications do not have to be installed on the remote computer nor do they have to be connected to a local area network to function. The application is run from the cloud and data generated are stored on the cloud, which can be accessed anytime by anyone with Internet service and proper access credentials. While many organizations see these as extended capabilities, it also has increased risk.

Both Microsoft® and Google® are providing organizations with free email, as well as online communications, applications, and storage. Many organizations are using these services of the cloud-based networks because of low cost and limited personnel needed while providing an enhanced service to the organization. There are risks associated with movement to cloud-based email systems. While it is easy to monitor servers when they are run by internal data centers and under the control of the organizations IT department, it is more difficult when there is little control over servers that are located somewhere in the cloud. Therefore, it is important to measure and analyze not only performance, but also security of hosted information. It is very important that organizational leadership understand this and know how to respond to risks.

Internet Browsing

Internet browsing provides the means to explore information on worldwide computer networks, usually by using a browser such as Microsoft Internet Explorer, Google Chrome, or Firefox. The browser allows access to rich educational and cultural resources (text, sounds, pictures, and video). This also gives users an improved ability to understand and evaluate information and stay informed by accessing Web sites.

Risks associated with Internet browsing relate to sites with inaccurate, misleading, and untrue information. There also is access to sexually explicit images and other sites promoting hatred, bigotry, violence, drugs, cults, and other things not appropriate in the organizational setting. The Internet, in general, has no restrictions on marketing. Some Internet sites deceptively collect personal information in order to sell products via requests for personal information for contests and surveys. The Internet is a relatively wide-open interface to share data without any form of censorship.

Online Chatting

Online chatting is a popular communications tool used by many. Online chatting is reading messages from others as they are typing them, usually in a theme- or social network-specific interface. The inclusion of chatting in social networking sites has helped to maintain its appeal as a widely used communications medium. Chatting is popular because it allows users to communicate with people from around the world by synchronous typing of text into a chat interface. Staff connecting to others via Web sites or social networking portals, in itself, does provide inherent risk.

Chatting is a risky environment because it provides an interface where people can communicate, in real time, with as much anonymity as they desire. Social networking sites add a new dimension to chatting because of the ability to have an online profile. Online profiles make searching differing demographic groups easy. Hackers will befriend users via social networking sites and gain trust by behaving as an understanding and trusted friend. Once trust is gained in the chat environments, staff can be susceptible to illicit activities.

SOCIAL MEDIA

The rise of Web 2.0 and, most notably, online social media has had a profound effect on society and organizations as well. Web 2.0 combined with broadband Internet access has changed the way people communicate, process, and store data. The diffusion of Web 2.0 has impacted users across all demographic groups. Its technologies include social networking sites, blogs, wikis, video-sharing sites, hosted services, Web applications, and tags. For users, these tools, which are typically free or low-cost, represent a transition from institutionally provided to freely available technology.

Web 2.0 technologies are possible because of the adaption of programming tools, such as asynchronous JavaScript (AJAX). AJAX is a group of interrelated Web development tools used to program interactive Web applications. This programming aspect, combined with diffusion of residential broadband Internet access, created an environment that evolved and migrated users to a very interactive form of the World Wide Web that facilitated interactive social networking. This AJAX-infused adaptation of the World Wide Web is simply referred to as Web 2.0.

Web 2.0 is the current rendition of the World Wide Web that provided a "social" approach to generating and distributing Web content, characterized by open communication, decentralization of authority, and freedom to share and reuse information (Acar, 2008; Madge et al., 2009; Subrahmanyam et al., 2008). The socialization through Web 2.0 interaction is exemplified by photo and video sharing (Photobucket.com, Flicker. com, and youtube.com), wikis and blogs, (Wikipedia.com, TripAdvisor.com, and UrbanSpoon.com) as well social networking sites (Facebook.com and Twitter).

Online networking has proliferated through the Web 2.0 environment because of an interactive design attribute that has been further propagated by increased residential Internet access and bandwidth (Acar, 2008; Madge et al., 2009; Subrahmanyam et al., 2008). The World Wide Web today has developed into a network of participation that typifies online social networking. There are a variety of online social networks in existence, the most popular being Facebook and Twitter.

Social networking sites combine common characteristics that allow profile creation, friends listing, and public viewing of friend lists. Online social networking Web sites also allow users to create their unique Web presence referred to as their *profile*. Through the profile, the social network users assume false identities while exploring friendships and relationships with other individuals who also have profiles on that Web site. Online social networking does not function entirely in real time, like conventional chat rooms and instant messaging, so the interactions that take place are not always instantaneous even though most have chat room functionality.

Most social networking profiles are developed from responses to questions that request a user to disclose a variety of personal information (Steinfield, Ellison, and Lampe, 2008; Mazer, Murphy, and Simonds, 2007; Mitrano, 2008). Personal information includes user names or other identifiers, such as sexual preference, organizations, geographical location, and the extent of the relationship that the persons are currently in or seeking with others. The profile also allows for self expression through personal photographs and videos. Persons also can make available their list of friends, member groups, as well as create an area where individuals can post remarks or statements from others. The individual's social networking profile has distinct Web addresses that can be bookmarked or linked allowing others to use and share that data with third parties.

The most popular online social network is Facebook, an online directory that connects people through social networks. The Web site www.facebook.com (Facebook) was initially designed and developed in 2004 by Mark Zuckerburg, a Harvard sophomore, and was inspired by a widely known paper version of a college facebook. The directory consisted of individuals' photographs and names, and was distributed at the start of the academic year by university administrations with the intention of helping employees get to know one another.

Zuckerburg's initial intention was to create an online Web site to help Harvard co-eds get to know one another for the purpose of finding roommates (Shier, 2005). It is no surprise that the Facebook Web site has grown in popularity among personnel because it was initially designed exclusively for personnel.

From its creation in 2004, Facebook has grown from hundreds of users to more than 1 billion active users (Facebook, 2011). Facebook was originally designed for college staff, but now is open to anyone 13 years of age or older. In 2006, Facebook lifted its .edu requisites, where users had to have an email address with an education suffix (i.e., an email address ending in .edu). With that development, there was a mass movement to Facebook as the social networking portal of choice for most social network users (Mazer, Murphy, and Simonds, 2007; 2009).

The Web 2.0 aspect of Facebook provides a tool for friends to keep in touch and for individuals to have a presence on the Web without needing to build a Web site. Facebook makes it easy to upload pictures and videos, making its use so simple that nearly anyone can publish a multimedia profile (Mitrano, 2008). Facebook has made it easy to find friends using an email address or to search by name or pull up listings based on a variety of demographic variables. With a public profile on Facebook, a person can be found by the other 500 million users.

Each Facebook profile has a "wall" where friends can post comments. Because the wall is viewable by all the person's friends, wall postings are basically a public conversation. By default, people can write personal messages on friends' walls or send a person a private message that will show up in their private Inbox similar to an email message. Facebook offers tools to develop and maintain relationships that are of particular importance to those emerging into adulthood. Recently, the use of messaging via online social networking has surpassed emails as the primary means of communication between staffs (Smith, 2010).

Facebook allows each user to set privacy settings. For example, if a user has not added a certain person as a friend, the user can have his/her privacy set so that other users will not be able to view the user's profile. A user can adjust the privacy settings to allow other users or peers to view portions or all of the profile. Users also can create a limited profile, which allows users to hide certain parts of the profile from a list of users that an individual selects.

Another feature of Facebook that users like is the ability to add gaming applications to a user page. Facebook applications are programs developed specifically for Facebook profiles. The most popular of these programs include interactive games, such as Farm Town, Mafia Wars, and thousands of other interactive multiuser applications. Because most game applications save scores or assets, friends can compete against each other or against millions of other Facebook users.

Providing a cyber-secure environment in organizations is ever challenging, not only because of the dangers inherent in the Internet and social networking, but the added dimension of users bringing their personal technology into the organizational environment that can access a host of technological resources. Users increasingly utilize their personal technology, most notably their smartphones, to access social networking sites.

Smartphones

The use of smartphones exceeds the use of traditional computers in accessing the Internet (Weintraub, 2011; Albanesius, 2011). A smartphone is a cellular phone that combines the functions of a networked computing system and a mobile phone. The smartphone hosts a wealth of technology and applications (apps) that are as robust as a computer in many situations. In addition to the standard audio and text capabilities, smartphones typically serve as a video/still camera, audio recorder, media player, and mobile computer. They incorporate apps and browsers that can do a number of things ranging from video conferencing, GPS navigation, and social networking via Wi-Fi and broadband Internet Protocol network access.

A growing issue with cyber life and smartphone usage relates to geo-location tools such as Facebook Places. These services have become very popular apps, especially in the social networking environment. The apps allow users to "check in" to locations (neighborhood businesses) via their mobile phone. Their location is then sent to their friends and in many cases with a map showing their exact location.

Geo-locating is used primarily as a marketing tool for businesses, giving reduced cost or free merchandise for app users who visit their store. However, there are some obvious security considerations. Every time someone checks into a location publicly, they are telling the world exactly where they are. Location sharing could encourage stalking, as well a host of other hazardous issues, because the user is broadcasting his/her physical location via the Internet or social media site.

Texting also needs to be considered when looking at smartphones and cybersecurity. Technically, texting is not a Web-based technology. However, it has become widely diffused and has become synonymous with modern youth in the proliferation of cell phone technology.

As most know, text messaging has become the preferred method of communication for many. While texting, like other cyber technologies has many positive benefits such as enhanced communication for users, it also can create issues for the security manager.

This section has attempted to provide as understanding of cyber use by personnel with both positive attributes and consequent risks. Providing a technology-secure environment is a very broad topic and includes many aspects of staff life. It also is obvious that technology will continue to play a large part in organizational environments and in the lives of staff members. Therefore, security managers should understand the potential role organizations can have in facilitating a secure cyber environment.

PROVIDING A SECURE CYBER ENVIRONMENT

The second part of this chapter addresses the role of providing a cyber-secure environment for staff members. Providing cybersecurity is contingent upon security managers understanding the environment, risks, and vulnerabilities of technology use. Because of the breadth of the issues and complexity of the environment, there is no more valid approach to achieving a secure cyber environment in organizations than to provide policy guidance to organizational stakeholders. This approach will use the security manager to facilitate safe, secure, and ethically sound practices by staff.

It is not the intent of this section to address cybersecurity from the view of a network administrator, but to make the security manager aware of what is needed to make an organization's cyber environment secure. An organization must have both policies and procedures available to provide for authentication, firewalls, and virus protection on its computer systems. Authentication will provide secure login and allow individuals to access only the data and applications they require. The firewall also serves as an access control device that will limit the user access to unwanted programs and data. Virus protection programs will protect the organization from viruses and malware on the network. Each of these three areas must be addressed by computing professionals in that environment. The best application for the security manager is to provide stakeholders with information regarding the risks associated with the cyber environment and, subsequently, how to make them safe, secure, and ethical employees.

The task of cybersecurity is daunting because there are so many aspects of a staff member's cyber life that are out of reach of the organization. Members have technology and they use it frequently outside the realm of organizational mentorship. They use technology to communicate peer-to-peer and often without knowledge, concern, or ethical understanding. Therefore, policies, training, and codes of conduct infused with cybersecurity, safety, and ethics are important to provide the company's personnel with the ability to make good cyber decisions. Teaching staff members to be good cyber citizens will better ensure they do not become cybercrime victims, or perpetrators, in the future. Because technology is diffused at all organizational levels, cybersecurity should be mandated throughout the organization.

One basic principle of cybersecurity is that staff members should be taught to avoid strangers in the cyber world. Staff must understand that people are not always who they say they are in cyberspace. If a staff member receives emails or other communications that seem suspicious, he/she should not respond, but make security personnel aware of it. Another basic principle that staff should understand is that they should not access nonwork-related content on the Internet. Be it audio, video, or some other media, it should be stressed in organizational policy that this is not allowed. Personnel must understand the concept of organizational cyber citizenship. At any level of technological interface, staff should be aware, as an underlying premise, that they must be good cyber citizens.

CYBERSECURITY IN THE ORGANIZATION

Making staff members into good cyber citizens is a goal for all organizational members. Computers and cyber technologies enhance staff performance by engaging, involving, empowering, and having security management address cybersecurity as a part of that relationship is essential. Moreover, it is the responsibility of organizations to provide basic cybersecurity for the staff, and ethical skills as a portion of their organizational process. Educate staff regarding an acceptable use policy and include the consequences for violating it. Instill a sense of cyber responsibility. The role of an organization in facilitating cyber citizenship is going to be even more challenged in the future. The best time to begin establishing a culture of cybersecurity in an organization is the first day you introduce your staff to technology.

EXERCISES

Many employees are using smartphones to access social networking sites while at work. They are using these forums to tease and taunt other employees, This, in effect, has created a disturbance in the organization.

- Does the issue of cyber threats differ from other organizational threats?
- What factors must be considered in addressing the issue?
- How should the security manager address the issue?
  1. Expand on this statement: The Internet is a paradox for many organizations.
  2. What does it mean to be "a good cyber citizen?"

REFERENCES

Acar, A. 2008. Antecedents and consequences of online social networking behavior: The case of Facebook. *Journal of Website Promotion* 3 (1/2): 62–83.

Albanesius, C. 2011. PCMag.com. Smartphone shipments surpass PCs for first time. What's next? Online at: http://www.pcmag.com/article2/0,2817,2379665,00.asp

Department of Homeland Security. n.d. Cybersecurity. Online at: http://www.dhs.gov/cybersecurity-overview

Key Facts. Facebook Newsroom. Facebook (retrieved June 23, 2013).

Madge, C., J. Meek, J. Wellens, and T. Hooley. 2009. Facebook, social integration and informal learning at university: It is more for socializing and talking to friends about work than for actually doing work. *Learning, Media and Technology* 34 (2): 141–155.

Mazer, J. P., R. E. Murphy, and C. J. Simonds. 2007. I'll see you on "Facebook": The effects of computer-mediated teacher self-disclosure on staff motivation, affective learning, and classroom climate. *Communication Education* 56 (1): 1–17.

Mazer, J. P., R. E. Murphy, and C. J. Simonds. 2009. The effects of teacher self-disclosure via Facebook on teacher credibility. *Learning, Media, & Technology* 34 (2): 175–183.

Mitrano, T. 2008. Facebook 2.0. *Educause Review* 43 (2): 72.

Rainie, L. 2011. The new education ecology. Paper presented at the 19th Annual Sloan Consortium, Orlando, FL, November 2011. Online at: http://www.slideshare.net/PewInternet/the-new-education-ecology

Shier, M. T. 2005. The way technology changes, how we do what we do. *New Directions for Staff Services* 1 (12): 77–87.

Smith, S. D. and Caruso, J. B. 2010. ECAR study of undergraduate students and information technology. Retrieved January 12, 2011 from http://www.educause.edu/library/resources/ecar-study-undergraduate-students-and-information-technology-2010

Steinfield, C., N. B. Ellison, and C. Lampe. 2008. Social capital, self-esteem, and use of online social network sites: A longitudinal analysis. *Journal of Applied Developmental Psychology* 29 (6) Nov.–Dec.: 434–445.

Subrahmanyam, K., S. M. Reich, N. Waechter, and G. Epinoza. 2008. Online and offline social networks: Use of social networking sites by emerging adults. *Journal of Applied Developmental Psychology* 29 (6) Nov.–Dec.: 420–433.

Weintraub, S. 2011. CNN Money. Industry first: Smartphones pass PCs in sales. Online at: http//tech.fortune.cnn.com/2011/02/07idc-smartphone-shipment-numbers-passed-pc-in-q4-2010/

12 Security Investigations

Investigation is the process for obtaining facts and information for a better understanding of an event or to determine the cause of an event. Security investigations involve fact finding for the purpose of identifying security breaches, loss, and safety concerns. An investigation may expose activity violating company policy, criminal laws, and administrative regulations. For example, an employee falsifying company records may be disciplined by the company and be subject to criminal penalties. The company may face a civil lawsuit or administrative sanctions for the activity.

The investigative process may be used to gather evidence in the event of a lawsuit. For example, an employee driving a company car to the post office collides with another vehicle causing injury to the driver of the other vehicle. The police collision report is designed to provide a neutral investigation and identify facts as to which party was at fault. The efficacy of the collision report can be biased by a number of factors: the officer omits a fact not important to his investigation, evidence may have been moved or overlooked, or information may have been incorrectly recorded. Investigation by company personnel seeks to preserve evidence, statements, and facts for future litigation against the company. As an FBI agent, the author investigated a fender-bender involving a citizen who was rear-ended by a government vehicle. The "severe" injury to the citizen's neck seemed out of place for the minor damage noted to citizen's vehicle; the government vehicle was not damaged. When questioned about his injury, the citizen stated that he would likely have to see his brother-in-law, the chiropractor, for the next two years. Realizing the mistake, he quickly added that he meant the statement as a joke. Nonetheless, the statement was recorded and used in later settlement negotiations.

Investigation may uncover evidence of a criminal nature. As the security investigation develops, information relating to a criminal act should be reported to the police. Some jurisdictions make it a criminal act in itself *not* to report a serious crime. Misprision of a Felony, 18 U.S.C. § 4, makes it a crime not to report a federal violation to the authorities. When law enforcement agencies are notified, they will take over the investigation … for a legitimate reason. They are responsible for enforcing the laws and must preserve evidence in preparation for criminal legal proceedings. Your cooperation with law enforcement will complement the investigation; however, it also may expose the company to civil and even criminal liability. Legal counsel should be informed of any requests for information, especially court orders. Failure to cooperate may result in criminal or court sanctions for obstruction of justice.

WHO, WHAT, WHEN, WHERE, HOW, AND WHY

Common questions to be answered by who, what, when, where, how, and why include:

Who committed the act? Who reported the event? Who has information on the event? Who witnessed the event?

What happened? What proof do we have that the event actually occurred? Is what happened a crime? Did what happened violate a policy? Could what happened subject us to a lawsuit?

When did the event occur? How do we know when the event occurred?

Where did the event occur? Was the event on property owned, leased, or controlled by the company? Does it matter for the company (for liability) where the event occurred?

How did this occur? How can we prevent it from happening again? How did we learn about the event? How did we respond to the event?

Why did this occur? Why did they commit the act? Why didn't they report the incident?

Security investigations are similar to, if not exactly like, criminal investigations. The basic questions of who, what, where, when, how, and why provide the foundation for solving the event. The more detailed answers to these questions lead to the identification of the problem and finding solutions to alleviate future events.

The investigation gathers evidence. Evidence is any object, statement, or thing that tends to prove or disprove a material fact. The investigator gathers evidence two ways: (1) personal observation and (2) the observation of others. In most cases, an event requiring further investigation is brought to the attention of the investigator. Discovery may be by security patrols, employees, visitors, anonymous tips, surveillance, etc. The event may be observed visually, audibly, or through examination of documents, tests, or measuring devices. Have in mind the objective of your investigation: arrest and prosecution of the culprit, recovery of goods, discovering the leak, etc. Valuable time can be wasted by going in a direction that is of no value to your objective. You may be more concerned with the recovery of your company's stolen goods, while law enforcement resources may be best used in the prosecution of the criminal act.

On discovery of the event, physical evidence should be preserved and protected from contamination (intentional or unintentional) from bystanders and other curious and interested persons. Security personnel should remain at the event scene to assure the integrity of the evidence. Searching and recovery of evidence should be conducted immediately, with photographs taken of each item. A diagram of the scene and location of evidence should be completed.

Crime scene diagrams should accurately depict the scene of the event. The room or area and locations of furniture, evidence, etc. should be measured and recorded to show the relationship of items at a scene. Before any evidence is moved, a photo should be taken, first with the item only, then with an identifying number or mark. This number should track with the evidence throughout the investigation.

Evidence should be stored in a secure area with a notation of the location, date, and who seized the evidence. Examination of evidence should be limited to persons necessary for the investigation, and all persons who have handled the evidence should be noted. Some evidence may need to be examined by a laboratory or other people with specialized experience. For example, financial records on a suspected embezzlement should be examined by a Certified Public Accountant with forensic accounting credentials. Evidence that is moved for examination must identify a chain of custody, again, to maintain the integrity of the evidence. Each person who has "custody" of the item, for any period of time, however brief, should be noted. If the need arises for the evidence to be introduced at trial, the chain of custody will prove that the item introduced at trial is the same item discovered at the event scene.

Witnesses should be interviewed and asked to submit a written statement. The basic who, what, when, where, how, and why should be asked to develop facts and generate leads for more information. A search for additional witnesses should fan out from the scene. In addition to questioning people, look for monitoring cameras that may have recorded the event and potential approach and escape routes from the event. In extreme cases, you may need to post signs or advertise for possible witnesses to the event. Rewards may encourage witnesses to come forward. Anonymous witness information can be effective in an investigation. While the witness's credibility cannot be verified, the information given can corroborated with known information, thereby adding to its credibility. Anonymous sources require further investigation, but can lead to the right person for answers and a solution to your event. Anonymous sources may lead you to an event before something occurs, allowing you to set up personnel or cameras to capture the culprit in the process of his/her corrupt act.

Investigators should be trained in interviewing skills to listen and watch for deceptive or revealing behavior from witnesses. It is not uncommon for the perpetrator to be among those who witnessed the event. It is good to have two investigators during interviews when possible. One investigator can conduct the interview and watch the witness for nonverbal behavior and listen to the words used, while the second investigator can focus on taking notes.

Interview notes are taken at the time of the interview. The purpose of the notes is to immediately record the witness's statement, paying attention to key facts that may have a bearing on the case. At the scene of an event, investigators work quickly to get an idea of what happened. Notes are often sketchy and written in a modified shorthand or in abbreviations or acronyms only the interviewer will understand when he/she prepares an interview report.

When a formal statement or report of interview is written, it is possible for interviewer bias that will omit a point that does not support his/her position. If the formal statement is challenged, the notes are used to clarify the statement. Since the notes were prepared at the time of the interview, which is closer in time to the event than when the report of interview was prepared, they are more likely to accurately reflect the event.

Interviews of potential subjects should always be conducted with two investigators, primarily for the reasons listed above. Two investigators complement each other; one investigator may think of a question the other didn't think of. Subject interviews are usually conducted in an office or interview room. If possible, the subject interview should occur in the security office. The room should be free from distractions. A picture on a wall can give the interviewee something on which to focus. Chairs should be positioned with no table between the investigators and the interviewee. This allows you to watch for body language. (??) liked to place the interviewee in a swivel chair. The chair would telegraph movements, which can indicate nervousness, deception, or both.

At the conclusion of the interview, a statement should be prepared by the investigators for the interviewee to sign, especially when the interviewee confesses to the act or makes admissions indicating some culpability in the matter. A confession is an admission of guilt that accounts for all of the elements of a crime or civil offense. Admissions are when a person admits to an act, or part of an act, but not culpability for a crime or civil offense. Admissions can be used to prove guilt, e.g., persons may admit they took a computer from the supply room, but deny that they stole it.

The investigator should prepare the statement rather than have the interviewee write out a personal statement. The interviewee may intentionally or inadvertently leave out key information. The interviewee is given the statement and time to read it and make corrections before acknowledging the statement as a true reflection of his or her comments. Interview notes are always retained.

A polygraph exam may be conducted to assess the veracity of witness or suspect statements. Polygraph exams are administered by trained and licensed examiners who read the results of physiological data collected from the examinee. The examiner uses the results to form an opinion on the truthfulness of the examinee. The Employee Polygraph Protection Act limits employers' use of polygraph exams. Generally, covered employers cannot require or request that an employee take a polygraph exam. Exemptions from the act are government agencies, and some security related firms, such as an alarm or guard service.

All evidence, records, and statements should be stored in an investigation file. A filing system should be established so files can be found and accessed efficiently. The file should identify the location of evidence and list the status of the case. An investigative report should summarize all of the evidence and list potential witnesses and their potential testimony. A complete history of suspects should be included as well.

THE INVESTIGATIVE REPORT

When all of the evidence has been collected and examined, statements taken, and reports received, the case file should be reviewed and material prepared and included into a report. (See below for a format for investigative reports.) The investigative report begins with a statement of the event and persons responsible or liable for the event. Provide a synopsis of the matter so that the reader can get the general idea of what happened and what action you are proposing. Your action may suggest criminal prosecution; civil remedies that may include employee disciplinary action, changes to policy, or preparation from civil action against the organization.

A Table of Contents page helps the reader move quickly to key areas in the report. The Appendix will contain detailed reports and other documents useful for prosecution or civil action, or defense from prosecution or civil action. In between the Table of Contents and Appendix are summaries listed according to key evidence.

Because this document may go outside the organization and even become public in a trial, trade secrets and other proprietary information should be noted, but not included in your report. The caution here is not to try and conceal or misinterpret evidence. All evidence should be noted, with confidential material being identified as such. Attorneys will deal with disclosure in deciding to prosecute a matter.

FORMAT FOR INVESTIGATIVE REPORT

- Cover Page:
    - Title of investigation
    - Case number
    - Name of subjects
    - List of violations
    - A synopsis of acts and events
- Table of Contents: (For headings delineated below)
- Criminal/Civil/Administrative History:
    - Related proceedings
    - Contact persons
- Potential Witnesses:
    - Name of witness and what they will testify to
    - Page number where the witness statement will be found
- List of Documents
- List of Evidence:
    - Who will testify to what piece of evidence
    - Location of evidence (police storage, company vault, etc.)
- Appendix:
    - Documents, statements, reports
    - Case studies
    - Security reports identifying open doors; access control shows employee in office late hours and weekends; witness reports that employee commented on windfall to explain change in lifestyle.

13 Security Management for Chemicals and the Chemical Facility Antiterrorism Standards

DEPARTMENT OF HOMELAND SECURITY STANDARDS AND REQUIREMENTS

Chemicals pose a great risk to many organizations and a challenge to security managers. The concern of this chapter is not just with organizations that produce or store large chemical stockpiles, but also those facilities that use chemicals that are secondary to their primary mission. The focus of this chapter is going to be on the impact that the Department of Homeland Security (DHS) has had on the regulation and security required by those organizations that use and store chemical inventories. This chapter relies heavily on DHS documents to try and simplify the chemical regulation process and the subsequent role for security management. DHS provided its Chemical Facility Antiterrorism Standards as a framework for securing chemicals from the possibility of harm or theft by terrorist groups. DHS has the authority to regulate chemical facilities that present high levels of security risk.

On October 4, 2006, President George W. Bush signed the Department of Homeland Security Appropriations Act of 2007, which provides DHS with the authority to regulate the security of chemical facilities. The Chemical Facility Antiterrorism Standards (CFATS), 6 CFR Part 27 Interim Final Rule (IFR) was published on April 9, 2007 (DHS, 2007a). The purpose of 6 CFR Part 27 is to lower the risk posed by certain chemical facilities. CFATS requires chemical facilities to provide DHS with information to determine whether they are a covered facility and are required to meet certain security performance requirements. In order to identify high-risk chemical facilities, DHS has identified chemicals for preliminary screening based on the belief that such chemicals, if released, stolen, diverted, and/or contaminated have the potential to create significant human health and/or life consequences.

If a facility possesses a chemical that is on the DHS chemicals of interest list in 6 CFR Part 27, Appendix A, at or above the screening threshold quantity for any applicable security issue, the facility must complete and submit a chemical security assessment, know as Top-Screen, under DHS 6 CFR § 27.200 of CFATS. This authorizes DHS to collect information from chemical facilities on a broad range of topics related to the potential consequences of or vulnerabilities to an attack or incident. The CSAT Top-Screen is one method DHS may use to gather such information. The

Top-Screen is available through CSAT, a secure, Web-based system. After analyzing a facility's Top-Screen information, DHS will make a preliminary determination of whether a facility presents a high level of security risk and, therefore, must comply with additional requirements of CFATS (see Appendix C at the back of this book).

Organizations that have to adhere to this standard are extensive. DHS has attempted to identify facilities covered by its chemical security regulation. Facilities that may be required to comply with at least some provisions of the CFATS regulation will largely fall into the following categories:

- Chemical manufacturing, storage, and distribution
- Energy and utilities
- Agriculture and food
- Paints and coatings
- Explosives
- Mining
- Electronics
- Plastics
- Healthcare (DHS, 2007b)

To determine which chemical facilities meet the CFATS criteria for high-risk chemical facilities, the department developed the Chemical Security Assessment Tool (CSAT) Top-Screen, an easy-to-use, online questionnaire that must be completed by facilities that possessed any chemicals on the CFATS chemicals of interest list at or above the listed quantity for each chemical.

Exemptions to CFATS regulations include:

- Facilities regulated pursuant to the Maritime Transportation Safety Act (MTSA)
- Public water systems, as defined in the Safe Drinking Water Act
- Water treatment facilities, as defined in the Federal Water Pollution Control Act
- Facilities owned or operated by the Department of Defense or the Department of Energy
- Facilities subject to regulation by the Nuclear Regulatory Commission (DHS, 2007b)

The Department of Homeland Security does not currently regulate railroad facilities that are used to store large quantities of chemicals or materials in rail cars that are on the DHS chemical of interest list and does not request that railroads complete the CSAT Top-Screen. Likewise, the department has no intention at this time of requiring long-haul natural gas pipelines to complete the CSAT Top-Screen; however, chemical facilities otherwise covered by this regulation and with a pipeline within their boundaries must identify the pipeline as an asset and address it, as appropriate, in the Site Security Plan (DHS, 2007b).

Under CFATS, Congress directed the Department of Homeland Security to identify and secure those chemical facilities that present the greatest security risk. From their definition, security risk is a function of the following:

- The consequence of a successful attack on a facility
- The likelihood that an attack on a facility will be successful (vulnerability)
- The intent and capability of an adversary in respect to attacking a facility (threat).

Therefore, Congress and the administration have directed the department to ensure the security of specifically high-risk chemical facilities.

RISK-BASED PERFORMANCE STANDARDS

Because each chemical facility faces different security challenges, Congress explicitly directed the department to issue regulations establishing risk-based performance standards for security chemical facilities. Performance standards are particularly appropriate in a security context because they provide individual facilities the flexibility to address their unique security challenges. Using performance standards rather than prescriptive standards also helps to increase the overall security of the sector by varying the security practices used by different chemical facilities. Security measures that differ from facility to facility mean that each presents a new and unique problem for an adversary to solve.

RISK-BASED TIERING

The department has developed a risk-based tiering structure that will allow it to focus resources on the high-risk chemical facilities. To that end, the department will assign facilities to one of four risk-based tiers ranging from high (Tier 1) to low (Tier 4) risk (DHS, 2007b).

Assignment of tiers is based on an assessment of the potential consequences of a successful attack on assets associated with chemicals of interest. The Department of Homeland Security uses information submitted by facilities through the Top-Screen and security vulnerability assessment processes to identify a facility's risk, which is a function of the potential impacts of an attack, the likelihood that an attack on the facility would be successful, and the likelihood that such an attack would occur at the facility.

Facilities that complete the CSAT Top-Screen and do not meet the consequence thresholds do not need to comply with CFATS. DHS recognizes that facilities have dedicated and invested time, resources, and capital to identify vulnerabilities and improve overall security. Facilities will be able to make use of information from these improvements. Facilities also may leverage their existing security measures in working toward compliance with CFATS and, specifically, the risk-based performance standards.

DHS considers a variety of factors in determining the appropriate tier for each high-risk facility, including information about the public health and safety risk, as well as the presence of chemicals with a critical impact on the governance mission and the economy.

The security measures needed to satisfy the risk-based performance standards for each covered facility correspond to the security risks presented by the facility.

Accordingly, facilities that present a higher risk will be required to meet more rigorous risk-based performance standards.

THE TOP-SCREEN PROCESS

The first step for a facility to take in determining whether it is covered under CFATS is to review the exemptions. Unless a facility is exempt, it must review to determine whether it possesses any chemicals of interest at or above the listed screening threshold quantities. DHS has listed the security issue(s) associated with each chemical of interest; each chemical presents at least one security issue, and some chemicals present multiple security issues. Where there are multiple security issues associated with a chemical, a facility must complete and submit a Top-Screen if it meets or exceeds the screening threshold quantities for any of the applicable security issues. If a facility determines that it possesses a chemical of interest at or above any applicable screening threshold quantities, the facility must register with DHS for access to DHS CSAT1 and complete the Top-Screen survey application. Using the information submitted via the Top-Screen survey application, DHS will make a preliminary determination whether the facility presents a high level of security risk.

The CSAT Top-Screen follows a logical, two-step data collection process (DHS, 2007c):

1. Step one involves collecting basic facility identification information.
2. Step two involves collecting information about the chemicals a facility possesses, manufactures, processes, uses, stores, and/or distributes.

Questions cover the following security issues:

- Release-toxic, release-flammable, and release-explosive chemicals with the potential for impacts within and beyond a facility
- Theft of explosive/improvised explosive device precursor (Theft/Diversion-EXP/IEDP) chemicals, theft of weapon of mass effect (Theft/Diversion-WME) chemicals, and theft of chemical weapon/chemical weapon precursor (Theft/Diversion-CW/CWP) chemicals
- Sabotage/contamination chemicals
- Chemicals that are critical to Government Mission and National Economy

The Top-Screen also provides additional questions for each of the following facility types:

- Petroleum refining facilities
- Liquefied natural gas storage facilities (DHS, 2007c)

Upon completion of the Top-Screen, facilities will be presented with one of two outcomes: (1) a preliminary determination that a facility is not high risk or (2) that a facility is high risk, and then DHS will notify the facility of:

- Its preliminary status as a high-risk facility.
- Its preliminary placement in a risk-based tier.
- The specific chemical of interest and related security issues that need further analysis.

Security Vulnerability Assessments

If DHS makes a determination that a facility is high risk, they will require the facility to complete a security vulnerability assessment to identify and assess the security of a facility's critical assets in light of the security issues raised by its possession of the subject chemicals.

Following a facility's submission of the security vulnerability assessment and its analysis by DHS, DHS will either confirm that a facility is high risk or inform a facility that DHS no longer considers the facility to be high risk or subject to further regulation under CFATS. For facilities confirmed to be high risk, DHS will communicate the final facility tier determination and the facilities must develop and implement site security plans that satisfy the risk-based performance standards enumerated in 6 CFR § 27.230.

SITE SECURITY PLANS

The Site Security Plan must meet the following standards:

1. Address each vulnerability identified in the facility's Security Vulnerability Assessment, and identify and describe the security measures to address each such vulnerability.
2. Identify and describe how security measures selected by the facility will address the applicable risk-based performance standards and potential modes of terrorist attack including, as applicable, vehicle-borne explosive devices, water-borne explosive devices, ground assault, or other modes or potential modes identified by DHS.
3. Identify and describe how security measures selected and utilized by the facility will meet or exceed each applicable performance standard for the appropriate risk-based tier for the facility.
4. Specify other information the DHS assistant secretary deems necessary regarding chemical facility security. Accept that a covered facility must complete the Site Security Plan through the CSAT process, or through any other methodology or process identified or issued by the assistant secretary. Covered facilities must submit a Site Security Plan to the department in accordance with the schedule provided in § 27.210. When a covered facility updates, revises, or otherwise alters its Security Vulnerability Assessment pursuant to § 27.215(d), the covered facility shall make corresponding changes to its Site Security Plan. A covered facility also must update and revise its Site Security Plan in accordance with the schedule in § 27.210. A covered facility must conduct an annual audit of its compliance with its Site Security Plan.

RISK-BASED PERFORMANCE STANDARDS

Covered facilities must satisfy the performance standards identified in this section. The assistant secretary will issue guidance on the application of these standards to risk-based tiers of covered facilities, and the acceptable layering of measures used to meet these standards will vary by risk-based tier. Each covered facility must select, develop in their Site Security Plan, and implement appropriately risk-based measures designed to satisfy the following performance standards:

1. Restrict Area Perimeter: Secure and monitor the perimeter of the facility.
2. Secure Site Assets: Secure and monitor restricted areas or potentially critical targets within the facility.
3. Screen and Control Access: Control access to the facility and to restricted areas within the facility by screening and/or inspecting individuals and vehicles as they enter, including:
    a. Measures to deter the unauthorized introduction of dangerous substances and devices that may facilitate an attack or actions having serious negative consequences for the population surrounding the facility.
    b. Measures implementing a regularly updated identification system that checks the identification of facility personnel and other persons seeking access to the facility and that discourages abuse through established disciplinary measures.
4. Deter, Detect, and Delay: Deter, detect, and delay an attack, creating sufficient time between detection of an attack and the point at which the attack becomes successful, including measures to:
    c. Deter vehicles from penetrating the facility perimeter, gaining unauthorized access to restricted areas, or otherwise presenting a hazard to potentially critical targets.
    d. Deter attacks through visible, professional, well maintained security measures and systems, including security personnel, detection systems, barriers and barricades, and hardened or reduced value targets.
    e. Detect attacks at early stages, through countersurveillance, frustration of opportunity to observe potential targets, surveillance and sensing systems, and barriers and barricades.
    f. Delay an attack for a sufficient period of time so to allow appropriate response through onsite security response, barriers and barricades, hardened targets, and well-coordinated response planning.
5. Shipping, Receipt, and Storage: Secure and monitor the shipping, receipt, and storage of hazardous materials for the facility.
6. Theft and Diversion: Deter theft or diversion of potentially dangerous chemicals.
7. Sabotage: Deter insider sabotage.
8. Cyber: Deter cyber sabotage, including by preventing unauthorized onsite or remote access to critical process controls, such as Supervisory Control and Data Acquisition (SCADA) systems, Distributed Control Systems (DCS), Process Control Systems (PCS), Industrial Control

Systems (ICS), critical business systems, and other sensitive computerized systems.

9. Response: Develop and exercise an emergency plan to respond to security incidents internally and with assistance of local law enforcement and first responders.

10. Monitoring: Maintain effective monitoring, communications, and warning systems, including:

    g. Measures designed to ensure that security systems and equipment are in good working order and inspected, tested, calibrated, and otherwise maintained.

    h. Measures designed to regularly test security systems, note deficiencies, correct for detected deficiencies, and record results so that they are available for inspection by the department.

    i. Measures to allow the facility to promptly identify and respond to security system and equipment failures or malfunctions.

11. Training: Ensure proper security training, exercises, and drills of facility personnel.

12. Personnel Surety: Perform appropriate background checks on and ensure appropriate credentials for facility personnel, and, as appropriate, for unescorted visitors with access to restricted areas or critical assets, including:

    j. Measures designed to verify and validate identity.

    k. Measures designed to check criminal history.

    l. Measures designed to verify and validate legal authorization to work.

    m. Measures designed to identify people with terrorist ties.

13. Elevated Threats: Escalate the level of protective measures for periods of elevated threat.

14. Specific Threats, Vulnerabilities, or Risks: Address specific threats, vulnerabilities, or risks identified by the assistant secretary for the particular facility at issue.

15. Reporting of Significant Security Incidents: Report significant security incidents to the department and to local law enforcement officials.

16. Significant Security Incidents and Suspicious Activities: Identify, investigate, report, and maintain records of significant security incidents and suspicious activities in or near the site.

17. Officials and Organization: Establish official(s) and an organization responsible for security and for compliance with these standards.

18. Records: Maintain appropriate records.

19. Address any additional performance standards the assistant secretary may specify (DHS, 2007c).

Since 9-11, the security of chemicals have become heavily regulated by the DHS. Chemicals pose a great challenge to security managers. The concern of this chapter is not just with organizations that produce or store large chemical stockpiles, but also those facilities that use chemicals that are secondary to their primary mission. The impact that the DHS has on the regulation and security required by these organizations that use and store chemicals inventories is astounding.

EXERCISE

DHS and the CFATS regulations are concerned with mandatory security requirements for organizations that use and store chemicals. As part of this process, they must maintain standards of security in protecting these chemical assets. Provide a detailed list of those security requirements.

REFERENCES

Department of Homeland Security. 2007a. 6 CFR Part 27 Appendix to Chemical Facility Antiterrorism Standards, Final Rule. Online at: http://www.dhs.gov/xlibrary/assets/chemsec_appendixa-chemicalofinterestlist.pdf
Department of Homeland Security. 2007b. Identifying facilities covered by the chemical security regulation. Online at: http://www.dhs.gov/identifying-facilities-covered-chemical-security-regulation
Department of Homeland Security. 2007c. Chemical facility antiterrorism standards. Online at: http://www.dhs.gov/chemical-facility-anti-terrorism-standards

14 Energy Sector
Security Management

Security for organizations that are in the energy sector are susceptible to great risk and challenges for those responsible for maintaining their security. The concern of this chapter is not just with organizations that produce or store energy in their primary mission, but with the impact that the Department of Homeland Security (DHS) has had on regulations and security required by these energy sector organizations.

The energy sector represents a very important component of our nation's critical infrastructure. More than 80% of the country's energy infrastructure is owned by the private sector and is integral to growth and production across the nation (DHS, 2011).

The energy infrastructure is divided into three interrelated segments: electricity, petroleum, and natural gas. The U.S. electricity segment contains more than 6,413 power plants, including 3,273 traditional electric utilities and 1,738 nonutility power producers. Approximately 48% of electricity is produced by combusting coal, primarily transported by rail, 20% in nuclear power plants, and 22% by combusting natural gas (DHS, 2013).

The heavy reliance on pipelines to distribute products across the nation highlights the interdependencies between the energy and transportation sectors. The reliance of virtually all industries on electric power and fuels means that all sectors have some reliance on a secure and stable energy sector. Many energy sector stakeholders and security managers have extensive experience with infrastructure protection and, more recently, have focused their attention on cybersecurity.

PROTECTING THE ENERGY SECTOR

The energy sector is very vulnerable. In 2008, an analyst in the Central Intelligence Agency (CIA) stated publicly that cyber attacks had already been used to disrupt electrical power in multiple cities outside the United States (McMillan, 2008). This alarming report, and others, fails to describe the extent to which a complex system, such as the national electric grid, could be vulnerable to cyber attacks.

Demonstrations of a cyber attack were exemplified against control systems that regulate an electric generator in Idaho. In this demonstration, under the code name "Aurora," the Department of Energy Idaho National Laboratory manipulated the generator's controls to exploit system weaknesses that caused the generator to fail. In particular, the attack caused extreme vibrations, which, in turn, physically destroyed internal components and ultimately caused the generator to catch fire (Meserve, 2007). This kind of attack, which was demonstrated to show feasibility, is likely to be even more effective in much larger generators, such as those in big dams and many coal-fired power plants.

The huge task of protecting the energy sector organizations is much like security in other organizations. By using a methodical process, the security manager must understand the environment, including its assets, their vulnerabilities, and criticalities to formulate a means for protection.

IDENTIFY ORGANIZATIONAL ASSETS AND GOALS

The first step of the energy sector security management process is to identify assets and specific overall goals for the organization. This part of the planning process should include a detailed overview of each organizational asset and its relationship to goals. Where possible, objectives should be described in quantitative or qualitative terms, and these goals should be measurable. Being able to measure needs, as well as outcomes, is fundamental to security management as well as to the organization.

The application of energy sector security programs should follow the critical thinking process regardless of the asset. The asset will have a direct impact on the application of the process and how countermeasures are to be implemented. Application of this security standard ensures a comprehensive approach to meeting organizational security needs in the threat environment, and that the scope of security is commensurate with the risk posed to an asset, relative to cost.

The Energy Sector has identified six general asset or system characteristics that are important parameters for evaluating the vulnerabilities of its infrastructure and developing risk management programs (DHS, 2011):

- Physical and location attributes: This relates to the physical location of the asset and the role that geographic location plays in the asset's inherent presence.
- Cyber Attributes: This considers the impact that cyber systems play in relationship to the asset.
- Volumetric or throughput attributes: This attribute is related to the capacity usage of the asset.
- Temporal/load profile attributes: This attribute relates, for example, to the varying loads the system may utilize based on seasonal energy usage.
- Human attributes: This relates to the technical knowledge of the energy sectors workforce in maintaining the sector's safety, reliability, and security.
- Importance of asset or system to the energy network: This relates to the interdependency and interrelationships between the larger energy grids.

The security manager must understand the relationship of assets in organizational structures and functions. This includes the physical and logical relationships that assets have with each other in the organizational environment. It also can relate to the impact one asset can have on the other in a loss event. An essential tool for better understanding the relationship between organizational assets is through the configuration of an asset hierarchy due to the dependencies with and interdependencies among energy sector infrastructures.

Analyzing Risks to Organizational Assets

An effective security management system demonstrates a careful evaluation of how much security is needed to protect organizational assets. Security managers must realize that too little security means that organizational assets can be easily compromised, while too much security can make assets hard to use or so degraded that performance is negatively affected. Security must be inversely proportioned to an asset's utility. It is given that there is always going to be risk associated with assets and activities. The only way to completely eliminate the risk, in many cases, would make that asset inoperable. Therefore, the role of security management is to find the optimal relationship between organizational processes, assets, and functionality. While the risk assessment process is more deeply analyzed in Chapter 9, it still maintains merit when looking at the overall role in building a security management program.

If the organizational asset is viewed through the inventory of potential loss events, the security manager must recognize the findings are not necessarily all inclusive. For each undesirable event where the assessed risk is either less than or exceeds the baseline level of protection, the security manager must identify the countermeasures that will provide a level of protection equivalent to the level of risk. For lower-lever risk, minimum countermeasures are typically less stringent, but also may be less effective in mitigating higher risks, while, at the other extreme, very high countermeasures are typically more stringent and generally more effective.

As defined in the NIPP (National Infrastructure Protection Plan)-Energy Sector (DHS, 2011) base plan, risk is a measure of potential harm that encompasses threat, vulnerability, and consequence, i.e., where an asset's risk is a function of the likely consequences (C) of a disruption or successful attack; the likelihood of a disruption or attack on the asset, often referred to as the threat (T) to the asset or the asset's attractiveness; and the asset's vulnerability (V) to a disruption or attack. As discussed in the sections below, the Energy Sector uses a variety of approaches that apply this widely accepted risk management principle to assess risk.

Security and risk managers in the energy sector have extensive experience in development and application of methodologies for assessing facility and system risk and prioritizing assets to be protected. Such methodologies have been developed by a variety of sector security partners, including individual energy companies that own and operate Energy Sector assets, professional and trade associations, academic institutions, research centers, and DOE, as an integral part of meeting its longstanding responsibilities for safety and security and implementing its Critical Infrastructure Protection (CIP) program for the Energy Sector.

Because of the diversity of assets in the Energy Sector, many risk assessment methodologies are used. Some methodologies are tailored to a specific segment of the sector (i.e., electricity, oil, natural gas, or their system components), while others are used to assess risks at the system or sector level.

The Department of Energy, in cooperation with sector security partners, has undertaken programs to assess the risks of key energy infrastructure assets and to provide technology, tools, and expertise to other federal, state, and local organizations and the private sector.

These programs have involved establishing partnerships with infrastructure owners/operators, state and local governments, and a wide range of industry associations. Products include vulnerability and risk assessment-related methodologies, checklists, lessons learned, support for policy analysis, and guidelines for various types of assets. The Energy Sector also has worked closely with DHS in developing and transferring risk assessment methodologies. The sector has participated in DHS's Buffer Zone Protection Program (BZPP) and has worked with it to develop Risk Analysis and Management for Critical Asset Protection (RAMCAP) modules for petroleum refining and LNG (liquefied natural gas) facilities. Given the diversity of facilities in the Energy Sector and the wide range of methodologies being used successfully to assess risk, a "one size fits all" risk assessment solution is not appropriate.

The Energy Sector will consider such criteria through the Critical Infrastructure Protection Advisory Council (CIPAC) as the sector evaluates how best to move forward in terms of vulnerability and/or risk assessments that will support DHS's national risk analysis goals and to improve these methodologies.

The desired levels of protection of assets should be critically determined via a risk-based analytical process or risk assessment. The process will focus on risk as a measurement of potential harm or loss from an undesirable event. Understanding risk means understanding threats, vulnerabilities, and consequences. The level of risk is the combined measure of threats, vulnerabilities, and consequences posed to assets from specified loss events.

If the existing level of protection is insufficient, a determination must be made as to whether the necessary level of protection can be achieved; specifically, if the countermeasure can be physically implemented and whether the investment is cost-effective. Cost-effectiveness is based on the investment in the countermeasure versus the value of the asset. In some cases, investment in an expensive countermeasure may not be advisable because the life cycle of the asset has almost expired. Additionally, consideration should be given to whether other countermeasures may take priority for funding. Note that cost-effective is a different determination than cost-prohibitive. A countermeasure is cost-prohibitive if its cost exceeds available funding. Funding may exist for a countermeasure, but it may not be a sound financial decision to expend that money for little gain, making it not cost-effective.

ASSESSING CONSEQUENCES

The potential physical and cyber consequences of any incident, including terrorist attacks and natural or manmade disasters, is the first factor to be considered in risk assessment. In the context of the NIPP base plan, consequence is measured as the range of loss or damage that can be expected. The consequences that are considered for the national-level, comparative-risk assessment are based on the criteria set forth in HSPD-7 (Homeland Security Presidential Directive 7). These criteria can be divided into four main categories (DHS, 2011):

1. Human Impact: Effect on human life and physical well-being (e.g., fatalities, injuries).

2. Economic Impact: Direct and indirect effects on the economy (e.g., costs resulting from disruption of products or services, costs to respond to and recover from the disruption, costs to rebuild the asset, and long-term costs due to environmental damage).

3. Impact on Public Confidence: Effect on public morale and confidence in national economic and political institutions.

4. Impact on Government Capability: Effect on the government's ability to maintain order, deliver minimum essential public services, ensure public health and safety, and carry out national security-related missions. For example, one DOE Power Marketing Administration (PMA), the Bonneville Power Administration (BPA), has used the following screening criteria to identify its most critical facilities: economic security, national security, public health and safety, generation, and regional and national grid reliability.

An assessment of all categories of consequence may be beyond the capabilities available for a given risk analysis. Most Energy Sector assets are not associated with mass casualties, but may have economic and long-term health and safety implications if disrupted. However, the redundancy of system-critical facilities and overall system resilience minimize the potential for such consequences.

The complexity, diversity, and interconnectedness of the Energy Sector dictate the need for assessing consequences at many different levels of detail:

- Asset or facility level
- System, sector, and urban area level
- Regional and/or national level

These interdependencies may have national, regional, state, and/or local implications and are considered to be an essential element of a comprehensive examination of physical and cyber vulnerabilities.

ASSESSING THREATS

The Energy Sector has a broad view of threat analysis, encompassing natural events, criminal acts, insider threats, and foreign and domestic terrorism. Natural events are typically addressed as part of emergency response and business continuity planning.

In the context of risk assessment, the threat component of risk analysis is calculated based on the likelihood that an asset will be disrupted or attacked. Such information is essential for conducting meaningful vulnerability and risk assessments. Therefore, the Energy Sector strongly believes that relevant and timely threat information must be disseminated whenever possible. A number of sector representatives hold national security clearances that facilitate the sharing of classified threat information.

The DHS Homeland Infrastructure Threat and Risk Analysis Center (HITRAC), which conducts integrated threat analysis for all CI/KR (critical infrastructure and key resource) sectors, will work in partnership with owners and operators and other federal, state, and local government agencies to ensure that suitable threat information is made available. Furthermore, the same level of partnership must exist within all levels of federal, state, and local law enforcement.

The following types of threat products provided by HITRAC are needed for the Energy Sector:

- **Common Threat Scenarios**, which present methods and tactics that could be employed in attacks against the U.S. infrastructure.
- **General Threat Environment Assessments**, which are sector-specific threat products that include known terrorist threat information and long-term strategic assessments and trend analyses of the evolving threats to the sector's critical infrastructure.
- **Specific Threat Information**, which is critical infrastructure-specific information based on real-time intelligence and will drive short-term measures to mitigate risk.

More specifically, they help energy facilities, local law enforcement, and others to be more aware of potential indicators of terrorist and/or criminal activity.

ASSESSING VULNERABILITIES

Vulnerabilities are the characteristics of an asset, system, or network's design, location, security posture, process, or operation that render it susceptible to destruction, incapacitation, or exploitation by mechanical failures, natural hazards, terrorist attacks, or other malicious acts. Vulnerability assessments identify areas of weakness that could result in consequences of concern, taking into account intrinsic structural weaknesses, protective measures, resiliency, and redundancies.

Historically, the Energy Sector has been proactive in developing and applying vulnerability assessment methodologies tailored to its assets and systems. However, no single vulnerability tool or assessment methodology is universally applicable. Individual energy companies use assessment tools that are developed by professional and trade associations, federal organizations, government laboratories, and private sector firms. The number of tools in use is large, and the vast majority of significant facilities in the Energy Sector have already undergone assessments using one or more of these tools.

PRIORITIZE INFRASTRUCTURE

The Energy Sector is characterized by large networks as opposed to discrete assets. These networks are designed to operate with certain levels of reliability, even if portions of them (discrete components or assets) are out of service.

The importance of many of the individual components in the network is highly variable, depending upon location, time of day, day of the week, month of the year, and many other variables. What might be a critical asset on a Monday morning in January may not be critical on a Saturday afternoon in May.

Owners and operators of Energy Sector assets and networks have screening processes to identify internal priorities related to business conditions and supply/network reliability to help them ensure continuity of operations. From a grid perspective, the nation's oil and natural gas pipeline systems and electricity grid are designed and operated with built-in redundancy to ensure a certain degree of reliability and resiliency. Industry planning criteria assume that a local grid area can be operated even if one asset is out of service. In addition, during unforeseen events, the industry provides mutual aid to assist in emergency response and prompt restoration (see Chapter 5).

Regional planning groups for the oil and natural gas industry, and, historically, the Natural Environment Research Council (NERC) and regional reliability councils for the electricity industry, continuously evaluate network reliability. Their functions are well developed and understood, and the effectiveness of mutual aid agreements can be significantly affected by the nature of an event. Mutual aid partners also could be impacted by an event, and a utility might have to go outside the region to obtain aid.

DOE will continue to work in partnership with Energy Sector security partners to evaluate and support existing protective programs and to develop and support new programs that effectively reduce the vulnerability of critical energy assets. The overall strategy will focus on efforts that support the sector's goals to ensure continuity of energy services and business through reliable information sharing, effective physical and cybersecurity protection, and coordinated response capabilities.

The cornerstone of the overall strategy is partnership with all key stakeholders in the public and private sectors. This approach will continue to take full advantage of the extensive experience and expertise of sector partners and will ensure that repercussions of planned activities are carefully considered. Below, we outline the methods that Energy Sector partners use to assess, select, and implement cost-effective infrastructure protective programs, and highlight some of the existing cooperative efforts within the Energy Sector.

DEVELOP AND IMPLEMENT PROTECTIVE PROGRAMS

The Sector-Specific Plan identifies long-term technological solutions for protecting physical assets, energy control systems, and related cyber systems. Some activities in different phases may proceed simultaneously, where feasible, to expedite improvements in CIP.

Throughout the process, DOE will continue to work with security partners within the framework of the Energy Sector's goals that support its vision of a "robust, resilient energy infrastructure in which continuity of business and services is maintained through secure and reliable information sharing, effective risk management programs, coordinated response capabilities, and trusted relationships between public and private security partners at all levels of industry and government."

Implement the Security Programs

When the security manager completes the critical analysis via asset criticality and vulnerability studies, risk assessments, and cost benefit analysis, then he/she may implement the security measures that have been determined to best fit the asset risk. Implementation of new security programs are best accomplished through stages to make it easier for the organization to adapt to changes in the working environment. The security manager and organizational management should understand that there may be user resistance to security functions. It is recommended that staged implementation be performed starting with the most critical or vulnerable assets.

Assignments and Timelines

As the organization implements security measures, it must establish timelines for completing the associated tasks to complete them. This portion of the security implementation process should consider the abilities of staff members and the time necessary to realistically complete projects. The amount of preparation required to implement security measures may limit its immediate achievability. If the security measure is no capital cost, such as policy and procedural changes, or can be incorporated into a new project, the countermeasure often can be implemented immediately. When countermeasures require advance budgeting or coordination with outside vendors, implementation may be delayed.

Monitor for Compliance

Effective security management depends on adequate compliance monitoring. Most often, violations of security practices, whether intentional or unintentional, become more frequent and serious if not detected and acted on. Compliance monitoring is two primary activities: detecting security violations and responding to them.

The security manager should document the response to violations, and follow up immediately after noncompliance is detected. The organization should have a designated response group to deal with security violations. Members of the response group should have access to organizational leadership so that severe situations can be dealt with when needed.

A critical part of noncompliance should be the generation of reports for organizational leadership that discusses security violations. An additional objective of monitoring security measures for noncompliance is to identify potential security violations before they dilute the effectiveness of the program or before they cause serious damage.

Reevaluate Assets and Risks

Security management is a discipline that should be dynamic. As changes in the organization or assets occur, a reassessment of the security measures also should occur. Organizational leadership should keep security management abreast of larger changes in the organization so that security operations and measures are prepared to meet these challenges. The importance of sector assets is impacted by changing threats and continually changing consequences. Prioritization in the Energy Sector is dynamic—it changes constantly and goes on continuously. Static prioritization of Energy Sector assets could lead to critical decision making based on outdated or erroneous asset information in efforts to direct scarce resources to these assets, systems, and networks that may be the most critical at any point in time. DOE works with DHS to identify gaps in existing energy information and to identify publicly available databases or sources that could provide data to support DHS efforts to prioritize assets.

Some DHS, DOE, and other government programs need to allocate resources based on their prioritization (e.g., DHS's BZPP), Site Assistance Visits, and comprehensive reviews, as well as state and local initiatives). These programs supplement and support industry efforts. State and local efforts under the NIPP will be based on some measure of the relative importance, risk consequence, and vulnerability of the critical infrastructures within their jurisdictions. This will require that they work with the Energy Sectors in their jurisdictions so as to understand the importance of critical facilities. In addition, they will need to address policy, regulatory, or other barriers to undertake needed measures, and to allow for recovery of prudently incurred costs for those utilities subject to rate regulation. In addition DHS is providing funding to state and local entities based on risk assessments of critical infrastructures. The National Asset Database (NADB) also is organized by criteria that may not fully capture the relative importance of energy infrastructure from a systems perspective (DHS, 2011).

Developing a Security Management Program requires a broad field of knowledge in asset loss prevention, physical security, occupational safety, and intangible asset protection functions. It requires a comprehensive knowledge of organizational assets and the development and implementation of physical measures, policies, procedures, and guidelines to protect those assets.

Security management requires critical thinking skills in developing mechanisms to protect organizational assets. The process of security management utilizes processes of critical thinking providing a basis for a comprehensive security management program. The dynamic nature of organizations and environments require that the security response also be dynamic.

EXERCISE

The energy sector is very complex in securing because it has relationships with multiple organizations throughout the energy sector. Explain the complexity that is a result of interrelationships, and how losses in one area can affect other areas.

REFERENCES

Department of Homeland Security. 2011. National Infrastructure Protection Plan: Energy sector. Online at: http://www.dhs.gov/xlibrary/assets/nppd/nppd-ip-energy-sector-snap-shot-2011.pdf

Department of Homeland Security. 2013. Energy sector: Critical infrastructure. Online at: http://www.dhs.gov/energy-sector

McMillan, R. 2008. CIA says hackers have cut power grid. PCWorld, January 19. Online at: http://www.pcworld.com/article/141564/article.html

Meserve, J. 2007. Sources: Staged cyber attack reveals vulnerability in power grid. CNN, September 26. Online at: http://www.cnn.com/2007/US/09/26/power.at.risk/index.html

15 Post 9/11 Security

The new millennium brought a paradigm shift in the United States that completely changed the condition and understanding of the world and, most notably, our security. The United States, with the cold war over, domestic terrorism at an all-time low, and a thriving economy, achieved a feeling of accomplishment and content. Al Qaeda and other terrorist groups had attacked U.S. personnel and property overseas; the U.S.S. Cole in Yemen, soldiers in Somalia, and U.S. Embassy bombings in Tanzania and Kenya, again, was "over there," and we felt safe at home in this country. On September 11, 2001, that changed. The attacks on the World Trade Center in New York, the Pentagon in Washington, D. C., and the hijacked Flight 93 crash in Pennsylvania signaled to the people in the United States that they were *not* safe. The security that we had long maintained was gone.

In the United States, many things happened after 9/11 regarding security. There was a new paradigm of Americans to the world and there was a public acceptance, if not demand, for increased security in the homeland. In the years following 9/11, the awareness of threats against our communities required us to be proactive in identifying threats. The USA Patriot Act of 2002 and the reorganization of government under the Department of Homeland Security (DHS) was an answer to that concern. Facility security, awareness programs, and intelligence gathering and sharing are common components of the effort to secure the homeland.[*]

Under the authority of Homeland Security Presidential Directive 7 (HSPD-7), 18 critical infrastructure sectors were established. These sectors are recognized as areas crucial for the security of the country, and each sector is managed by a Sector-Specific Agency (SSA) that provides sector-level performance feedback to the DHS. In accordance with the National Infrastructure Protection Plan (NIPP), each SSA is responsible for developing and implementing a Sector-Specific Plan (SSP) (DHS, 2007a).

The U.S. national security critical sectors include:

1. Food and Agriculture
2. Banking and Finance
3. Dams
4. Chemical
5. Communications
6. Commercial facilities
7. Critical Manufacturing
8. Defense Industry Base

---

[*] Infrastructure Information Act of 2002 (CII Act) and the implementing regulations at 6 CFR Part 29 to critical infrastructure information that is not customarily in the public domain and is voluntarily submitted to the DHS.

9. Emergency Services
10. Energy
11. Government Facilities
12. Healthcare and Public Health
13. Information Technology
14. National Monuments and Icons
15. Nuclear Reactors, Materials, and Waste
16. Postal and Shipping
17. Transportation
18. Water

The events of 9/11 brought the recognition that the United States did very little sharing of information in regard to risks and threats, in either public or private organizations. The DHS recognized that each sector already has considerable data available to support a wide range of consequence, risk, and vulnerability assessments. Thus, part of the mission of DHS is to facilitate the sharing of data that are collected and used by owners, operators, trade associations, and a variety of industry organizations.

Likewise, DHS collects a wide variety of information, principally through the authorities of various federal agencies and at the state and local levels. DHS has established communication links between federal, state, and local government representatives and industry. During times of increased security posture or emergency situations, the best information sources are the trusted relationships between government and industry.

GOVERNMENT PROGRAMS

INFORMATION SHARING

In a joint effort, DHS has partnered with the Sector Coordinating Councils (SCCs) to develop Homeland Security Information Network (HSIN), an Internet-based communications system that enhances reporting and information sharing and allows industry participants to communicate securely with each other, with other industry sectors, and with government agencies. HSIN has agreements with both public and private sector leadership to share information via a restricted-access network, which is for key personnel to exchange information with the department during emergency situations. The site provides threat awareness and relevant security analyses and presentations (DHS, 2007b).

FOCUSED PROGRAMS

DHS also facilitated the development of endurance and protecting programs with both public and private sector organizations in partnership with DHS and other appropriate federal agencies. Programs will draw from effective practice already in use by industry and from national laboratory efforts. Specific programs will be designed to account for the significant interdependencies between them (Presti, 2012).

Establishing roles and responsibilities for implementation of new resiliency and protective measures and programs will present both a challenge and an opportunity. For example, in light of increasing cyber threats, DHS calls on both the public and the private sectors to work more closely together in arriving at a conclusion that would be both equitable and effective. Such an agreement would likely include real-time information sharing and the development of best practices to protect critical infrastructure (Presti, 2012).

SHARED GOALS FOR SECTOR SECURITY

DHS supports sector-specific goals for security. Each of the 18 sectors of the critical infrastructure is led by various groups, if not by a federal agency designated to coordinate goals to enhance security in that area. Goals often relate to response planning exercises, and also to facilitate loss plans that incorporate federal, state, and local law enforcement. DHS also supports goals to further enhance communication and coordination across the individual sectors as well as through all sectors. These include:

1. Food and Agriculture, which is directed by the U.S. Department of Agriculture and the Department of Health and Human Services' (HHS) Food and Drug Administration.
2. Banking and Finance, which is directed by the Department of the Treasury.
3. Dams, which are directed by the state Dam Safety Offices.
4. Chemicals: Since the majority of Chemical Sector facilities are privately owned, they require direct relationship with the Department of Homeland Security.
5. Communications: As most are of the private sector, owners and operators of the majority of communications infrastructure are the primary entities responsible for protecting sector infrastructure and assets.
6. Commercial facilities: The majority of the facilities in this sector are privately owned and operated, with minimal interaction with the federal government and other regulatory entities.
7. Critical Manufacturing: The majority of the organizations in this sector are privately owned and operated, with minimal interaction with the federal government and other regulatory entities.
8. Defense Industry Base, which is directed by the Department of Defense.
9. Emergency Services: They have dependencies and interdependencies with multiple critical infrastructure sectors and the National Response Framework's Emergency Support Functions, which supply elements for both the operation and protection of Emergency Services' assets.
10. Energy, which is directed by the Department of Energy.
11. Government Facilities is directed by the Federal Protective Service and the Department of Education for the education subsector.

12. Healthcare and Public Health, which is directed by the Department of Health and Human Services (HHS).
13. Information Technology: The majority of information technology assets are privately owned and require direct relationship with the Department of Homeland Security.
14. National Monuments and Icons, which are directed by the Department of the Interior.
15. Nuclear Reactors, Materials, and Waste, which are directed by the Nuclear Regulatory Commission.
16. Postal and Shipping, which is directed by the Postal and Shipping Government Coordinating Council.
17. Transportation, which is directed by the Transportation Security Administration.
18. Water, which is directed by the Environmental Protection Agency.

INFORMATION SHARING AND COMMUNICATION

DHS has improved information sharing. While information sharing still may not be optimally utilized, a security manager can get a better understanding of the risk environments now than prior to 9/11. Both the public and private sector need credible, timely, actionable information to ensure that appropriate security investments, programs, and decisions are made to protect organizational assets. DHS has attempted to build a system of trust in both public and private sector security partners to shared information.

DHS has provided many resources in assisting security managers to better deal with risk and subsequent loss. DHS has provided new methods or better explains existing methods that are acceptable to all stakeholders for collecting, protecting, and, as necessary, sharing sensitive data on the vulnerabilities of assets and the protective programs to address them. The private sector will be understandably cautious in providing information needed for vulnerability assessments and disclosing the results of assessments it has conducted, and may be equally cautious about providing specifics on ongoing and planned protective programs (DHS, 2007a).

CONCLUSION

The new millennium brought a paradigm shift in the United States affecting our own perceptions of security. The security manager is now in a position to have a better understanding of the risks, threats, vulnerabilities, and criticalities of assets than ever before. For security managers, there is nothing more practical than a methodical process of thinking. For managers, good thinking pays off while poor thinking causes problems, such as wasting resources and time. Critical thinking allows security managers to envision their duties in a logical process while focusing on making decisions and solving problems. It has been the intent of this chapter to help the security manager realize resources available for his/her support in securing organizational assets in the post-9/11 standard.

EXERCISE

This chapter discusses the various sectors of critical infrastructure in the post 9/11 world. An underlying premise relates to infrastructure interdependencies and risk considerations involving cross-sector analysis. How could the infrastructure linkages across sectors adversely affect the performance of other infrastructures in the event of loss to another sector.

REFERENCES

Department of Homeland Security. 2007a. Critical infrastructure sectors. Online at: http://www.dhs.gov/critical-infrastructure-sectors

Department of Homeland Security. 2007b. Homeland Security Information Network. Online at: http://www.dhs.gov/homeland-security-information-network

Presti, K. 2012. DHS secretary calls for public, private partnership to protect critical infrastructure, October 31, CRN. Online at: http://www.crn.com/news/security/240012631/dhs-secretary-calls-for-public-private-partnership-to-protect-critical-infrastructure.htm

Appendix A

DHS RISK ASSESSMENTS

This appendix provides acronyms for risk methodologies, programs, and other terms frequently used in DHS, along with a brief description of each.

ADRA: Air Domain Risk Assessment
Description: A Transportation Security Administration risk assessment methodology established to rank risk to generic United States air domain assets (airports, airplanes, navigation towers, general aviation, charter, etc.) from acts of terrorism.

C/ACAMS: Constellation/Automated Critical Asset Management System
Description: A system that provides a capability for State and local users to build and manage inventories of local infrastructures, conduct vulnerability assessments of those infrastructures, develop incident response plans, and build and generate a wide range of reports.

CARVER: Criticality, Accessibility, Recuperability, Vulnerability, Effect and Recognizability
Description: A mnemonic composed of the above terms, that when applied to security risk management, are used to characterize assets.

CIII: Critical Infrastructure Interdependencies Integrator
Description: A Monte Carlo simulation tool developed in conjunction with Argonne National Laboratory that measures the time and cost of asset recovery and restoration for critical infrastructure. This acronym is sometimes displayed as CI3.

CIKR CRM: Critical Infrastructure and Key Resources Common Risk Model
Description: An Office of Infrastructure Protection quantitative scenario risk assessment methodology designed to enable defensible cross-sector comparisons and comparisons of risk to combinations of infrastructure within a jurisdiction, sector, or attack type. The methodology is designed to support return-on-investment evaluations of potential risk management alternatives.

CIMS: Critical Infrastructure Modeling System
Description: A system created by the Idaho National Laboratory as a high level model designed to identify interdependencies that exist across multiple infrastructure sectors.

CIPDSS: Critical Infrastructure Protection Decision Support System
Description: An Office of Infrastructure Protection system for analysis of cross-sector critical infrastructure consequences.

CTMS: CREATE Terrorism Modeling System
Description: The Homeland Security Center for Risk and Economic Analysis of Terrorism Events (CREATE) methodology and software system for assessing risks of terrorism within the framework of economic analysis and structured decision making.

CTMS/TMS: CREATE Terrorism Modeling System/Terrorism Magnitude Scale
Description: A base-10 logarithmic scale, produced in the CTMS. It measures human, financial, and symbolic consequences of the scenarios the CTMS produces.

ECIP ASSESSMENT: Enhanced Critical Infrastructure Protection Assessment
Description: An assessment conducted to identify vulnerabilities and enhance security in collaboration with federal, state, local, and private sector stakeholders.

FAIT: Fast Analysis Infrastructure Tool
Description: Created by the National Infrastructure Analysis Center under DHS direction, the tool produces regional economic analysis, asset descriptions, and other important information on assets for the National Infrastructure Analysis Center's internal analysts.

HAZUS: Hazards United States
Description: Developed by the Federal Emergency Management Agency, a nationally applicable standardized methodology and software program that estimates potential losses from earthquakes, hurricane winds, and floods.

HIRA: Hazard Identification and Risk Assessment
Description: A DHS methodology designed to identify hazards and associated risk to persons, property, and structures.

JSERA: Joint Special Events Risk Assessment
Description: An Office of Infrastructure Protection risk assessment methodology that provides operationally relevant risk and risk-informed analysis capable of assisting partners with decision making related to special events.

MAST: Maritime Security and Strategic Toolkit
Description: A tool designed to enhance Area Maritime Security Plans and allow ports to better integrate their security efforts into the broader Urban Areas Security Initiative planning construct.

MD SHARPP: Mission, Demography, Symbolism, History, Accessibility, Recognizability, Population, and Proximity

Description: A mnemonic conceived as a numeric assessment in which each of the criteria are evaluated and combined to produce an overall score. In security risk management this score is typically applied to key assets.

MSRAM: Maritime Security Risk Assessment Model

Description: A United States Coast Guard model designed to identify and prioritize risks to ports, waterways, and associated facilities.

NEXT GENERATION ABEL: Next Generation Agent Based Economic Laboratory

Description: A high resolution macroeconomic model created by the Sandia National Laboratory that measures economic factors, feedbacks, and downstream effects of infrastructure interdependencies.

NMSRA: National Maritime Strategic Risk Assessment

Description: A United States Coast Guard all-mission risk assessment that informs budget and planning guidance.

NTSRA: National Transportation Sector Risk Assessment

Description: A Transportation Security Administration risk assessment established to evaluate threats, vulnerabilities, and consequences of selected terrorist attack scenarios, identify needs for more detailed analysis, inform planning decisions, and establish a baseline for other periodic analyses and analytical activities related to transportation security.

OCTAVE: Operationally Critical Threat, Asset, and Vulnerability Evaluation

Description: An information system analysis tool designed for large organizations and sponsored by the U.S. Department of Defense.

PAWSA: Ports and Waterways Safety Assessment

Description: A United States Coast Guard risk assessment methodology designed to identify major waterway safety hazards, estimate risk levels, and evaluate potential mitigation measures.

RAMCAP: Risk Analysis and Management for Critical Asset Protection

Description: A risk methodology that uses a common risk framework for owners and operators of the nation's critical infrastructure to assess terrorist risk to their own assets and systems.

RAPID: Risk Assessment Process for Informed Decision Making

Description: An Office of Risk Management and Analysis program aimed at developing a strategic-level process to gauge future risks across the full range of DHS responsibilities to inform the DHS's annual Planning,

Programming, Budgeting, and Execution cycle of resource allocation decisions.

RMAT: Risk Management Assessment Tool

Description: A Transportation Security Administration agent based model for analyzing and making decisions about risk reduction options based on threat, vulnerability, and consequence data.

RSAT: Risk Self Assessment Tool

Description: A tool, formerly known as the Vulnerability Identification Self Assessment Tool (VISAT), used within the Commercial Facilities Sector, public assembly subsector, to conduct risk assessments at individual venues. RSAT (sometimes displayed as R-SAT) assists owner-operators with the identification of vulnerabilities and risks and provides recommendations for improving a venue's overall security posture.

SEAR: Special Events Assessment Rating

Description: An Office of Operations Coordination effort to provide a single federal interagency resource to assess and categorize the risk to domestic special events that do not rise to the level of a National Special Security Event.

SHIELD: Strategic Hazard Identification Evaluation for Leadership Decisions

Description: Collaboration between the Office of National Capital Region Coordination and the Office of Risk Management and Analysis to create a regional risk management model.

SHIRA: Strategic Homeland Infrastructure Risk Assessment

Description: An annual collaborative process conducted in coordination with the infrastructure protection and intelligence communities to assess and analyze the risks to the Nation's critical infrastructure and key resource sectors from natural and manmade hazards.

SNJTK: Special Needs Jurisdiction Tool Kit

Description: A methodology developed by the Office for Domestic Preparedness (later the Office of Grants and Training, the functions of which have been reassigned to the Federal Emergency Management Agency) designed to address jurisdictions with special needs, or specifically, jurisdictions with unique and complex circumstances where it is necessary to compare relative risk levels across dissimilar assets and critical infrastructure.

SNJTK/CAF: Special Needs Jurisdiction Tool Kit / Critical Asset Factor

Description: A primary component of the SNJTK that represents characteristics of assets that would result in significant negative impact to the organization if an asset were lost.

STAR: Strategic Threat and Action Report

Description: A precursor to SHIRA that provided decision makers with a comparative assessment of risks to the Nation and the actions taken to manage those risks.

TRAGIS: Transportation Routing Analysis Geographic Information System

Description: A model created by the Oak Ridge National Laboratory used to illustrate highway, rail, and waterway routes across the Nation and to determine optimal routes for normal and abnormal states of infrastructure operation.

TRAM: Transit Risk Assessment Methodology

Description: A Federal Emergency Management Agency process that leverages past assessments of vulnerability with threat and consequence information to create a roadmap for making funding allocation decisions.

TRAVEL: Transportation Risk Assessment and Vulnerability Evaluation Tool

Description: A Transportation Security Administration tool that is used in facilitated, on-site assessments of transportation assets.

VISAT: Vulnerability Identification Self-Assessment Tool

Description: VISAT was renamed RSAT; for information please see RSAT entry above.

WISE: Water Infrastructure Simulation Environment

Description: A tool created by the Los Alamos National Laboratories that is similar to the IEISS, which studies the interdependency relationships within and between the water sectors in-depth and models the benefits of investment to this infrastructure.

Appendix B

EXAMPLE SECURITY RISK ASSESSMENT MEETING HIPAA GUIDELINES FOR A SMALL PHYSICIAN PRACTICE

A major security issue in the health care environment is the protection of information. The Health Insurance Portability and Accountability Act of 1996, commonly known as HIPAA, required privacy rules that regulate the use and disclosure of Protected Health Information (PHI) held by "covered entities" generally, health care clearinghouses, employer sponsored health plans, health insurers, and medical service providers that engage in certain transactions. By regulation, the Department of Health and Human Services extended the HIPAA privacy rule to independent contractors of covered entities who fit within the definition of "business associates".PHI is any information held by a covered entity which concerns health status, provision of health care, or payment for health care that can be linked to an individual. This is interpreted rather broadly and includes any part of an individual's medical record or payment history. The follow is an example of a HIPAA compliance risk assessment.

> "Administrative Data Standards and Related Requirements". 45 Federal Register 160.102 (3 December. 2002).
> ONLINE. Available:
> http://www.hhs.gov/ocr/privacy/hipaa/understanding/special/research/research.pdf.
> "The HIPAA Privacy Rule". 45 Federal Register 164.501 (3 December. 2002).
> ONLINE. Available:
> http://www.hhs.gov/ocr/privacy/hipaa/understanding/special/research/research.pdf.

SAMPLE HIPAA SECURITY RISK ASSESSMENT FOR A SMALL PHYSICIAN PRACTICE

Administrative, Physical, and Technical Safeguards
Breach Notification Rule

HOW TO USE THIS RISK ASSESSMENT

The following sample risk assessment provides you with a series of sample questions to help you prioritize the development and implementation of your HIPAA Security policies and procedures. While this risk assessment is fairly lengthy, remember that the risk assessment is required and it is critical to your compliance with the Security Rule. These sample questions cover Administrative, Physical, and Technical Safeguards, and the Breach Notification Rule, and are only representative of the issues you should address when assessing different aspects of your practice.

133

Keep your completed risk assessment documents in your HIPAA Security files and retain them in compliance with HIPAA document retention requirements.

HIPAA Security requires Covered Entities to protect against any reasonably anticipated threats or hazards to the security or integrity of electronic Protected Health Information ("ePHI") and to implement security measures sufficient to reduce risks and vulnerabilities to a reasonable and appropriate level. Assessing risks is only a first step. You must use the results of your risk assessment to develop and implement appropriate policies and procedures.

Reproduction and use of this form in a physician's office by physicians and their staff is permitted. Any other use, duplication, or distribution of this form requires the prior written approval of the American Physician Association. This form is educational only, does not constitute legal advice, and covers only federal, not state, law.

Sample Risk Analysis Directions

1. Review each of the following sample questions and rank the level of risk on a scale of 1 to 6 (with 1 being the lowest level of risk and 6 being the highest level of risk).
   a. "Risk for us" — 5 or 6 on your rating scale. You believe the situation or activity could put your practice at risk. For example, if your portable computer or smart phone contains scheduling or patient information, you have a high risk of exposing patient information if the ePHI it contains is not encryped.
   b. "Could be a risk" — 3 or 4 on your rating scale. For example, a poorly maintained inventory of electronic equipment would put you at risk of not being able to reconstruct your practice for an insurance claim in the event of a disaster.
   c. "Not a risk" — 1 or 2 on your rating scale. For example, the risk of flooding is likely to be lower for a physician practice that is located in a low flood risk area than for a practice located in a high flood risk area.
2. The following sample questions are designed to illustrate the kinds of questions that a physician practice should analyze in conducting its HIPAA Security Risk Analysis. Similar sample questions may appear in several sections because the sample questions correspond with various provisions of the Security Rule and are intended to allow you to think through your risks in different ways.
3. Identify a Security Official to develop and implement security policies and procedures and to oversee and protect confidential health information. For example, mobile computers often score a 5-6 risk rating. The Security Official should confirm that ePHI stored on such hardware is encrypted, and all such hardware is accounted for through either a check-in/check-out process or by storage in a locked cabinet at the end of each day.

Sample HIPAA Security Risk Assessment
For a Small Physician Practice

Implementation Specification	R/A	Sample Risk Assessment Question	Risk			Policy		Assigned to
			Risk for Us	Could be a Risk	Not a Risk	Policy in Place	Need Policy	
		Administrative Safeguards						
		Security Management Process 164.308(a)(1)						
		Team: Security Official, Physician, Workforce Members						
Risk Analysis	Required							
		Do you keep an updated inventory of hardware and software owned by the practice?						
		Can you identify where ePHI is located (e.g., desktops, laptops, handhelds, tablets, removable media, servers, etc.)?						
		Could you locate the inventory in a disaster (fire, flood, explosion, theft)?						
		Do you know the current approximate value of your hardware and software?						
		Does the inventory contain all necessary contact information, including information for workforce members and service providers?						
		Do you control the information contained on your information system?						
		Do you or your workforce take home portable computers or other devices containing ePHI?						

(continued)

Sample HIPAA Security Risk Assessment
For a Small Physician Practice

Implementation Specification	R/A	Sample Risk Assessment Question	Risk			Policy			Assigned to
			Risk for Us	Could be a Risk	Not a Risk	Policy in Place	Need Policy		
		Does any vendor have access to confidential patient data? Have you discussed HIPAA Security and HITECH requirements with such vendor(s)? Is an up-to-date Business Associate Agreement in place for each vendor that has access to ePHI?							
		Can a vendor change confidential patient data? If so, are you monitoring audit logs for such changes?							
Risk Management	Required								
		Do you update your workforce members' training each time you develop and implement new policies and procedures? Do you document initial and continuing training?							
		Have you set user access to ePHI? Does access correspond to job descriptions (clinical, administrative, billing)?							
		Do you monitor reports that identify persons and systems that access ePHI, including those not authorized to have access to ePHI?							
		Do you have control over who can amend your patient records?							

	Required								
Sanctions Policy		Have you developed a written sanctions policy against workforce members who do not abide by your policies?							
		Have you explained those sanctions to your workforce members?							
		Do you consistently enforce those sanctions?							
Information System Activity Review	Required								
		Do you regularly review system audit trails that identify who has accessed the system and track additions, deletions, or changes they may have made to ePHI?							
		Would you know if someone was trying to hack into your system? (Do you regularly review security incident reports?)							
Assigned Security Responsibility 164.308(a)(2) **Team: Physician, Security Official, Privacy Official**									
Assigned Security Responsibility	Required								
		Have you appointed a Security Official?							
		Do your Privacy and Security Officials coordinate privacy and security policies and procedures? (Privacy and Security Official may be the same person)							

(continued)

Sample HIPAA Security Risk Assessment
For a Small Physician Practice

Implementation Specification	R/A	Sample Risk Assessment Question	Risk			Policy		Assigned to
			Risk for Us	Could be a Risk	Not a Risk	Policy in Place	Need Policy	
		Workforce Security 164.308(a)(3) **Team: Security Official, Privacy Official**						
Authorization and/or Supervision	Addressable							
		Do you have written job descriptions that define appropriate access to ePHI?						
		Could an unauthorized workforce member obtain access to ePHI?						
		Are persons with access to ePHI supervised?						
Workforce Clearance Procedure	Addressable							
		Do you contact references before hiring employees?						
		Do you conduct background checks?						
Termination Procedures	Addressable							
		Do you immediately deactivate a workforce member's access upon termination (or, as appropriate, upon change of job description)?						
		Do you notify your IS vendor of an employee's termination within a specific time?						

Is there a standard checklist of action items when an employee leaves? (Return keys, close and payment of credit cards, return software and hardware)						
Does your practice consistently enforce checklists and policies with respect to all employees who are terminated or whose duties have changed, whether the termination or change was voluntary or for cause?						
Information Access Management 164.308(a)(4) **Team: Security Official, Physician**						
Isolating Health Care Clearinghouse Functions	Required					
If you use a health care clearinghouse that is part of a larger organization, have you confirmed that the clearinghouse has implemented policies and procedures to protect ePHI from unauthorized access by the larger organization?						
Access Authorization	Addressable					
Are you using your IT system's log-in process to authorize access (such as limiting administrative access)?						
Is each workforce member's access to ePHI based on his or her job description?						

(continued)

Sample HIPAA Security Risk Assessment
For a Small Physician Practice

Implementation Specification	R/A	Sample Risk Assessment Question	Risk			Policy		Assigned to
			Risk for Us	Could be a Risk	Not a Risk	Policy in Place	Need Policy	
Access Establishment and Modification	Addressable							
		Do you document, periodically review, and modify as appropriate workforce members' access to ePHI?						
Security Awareness and Training 164.308(a)(5) **Team: Security Official, with Input from Privacy Official**								
		Have you implemented a security awareness and training program for all members of your workforce, including management?						
Security Reminders	Addressable	Have there been lapses in privacy safeguards that indicate a need for training refreshers?						
		Have you identified your security training priorities?						
		Are security reminders posted in a visible location?						
		Are vendors aware of your security reminders?						
		Do workforce members know where to find a copy of your security policies and procedures?						
		Do workforce members understand the consequences of noncompliance with those policies?						

		Are workforce members with laptops, PDAs, or cell phones aware of encryption requirements?						
		Do you consistently follow your security awareness and training program with all new hires?						
Protection from Malicious Software	Addressable							
		Have you installed anti-virus and other anti-malware protection software on your computers? Do you use it to guard against, detect, and report any malicious software? Do you protect against spyware?						
		Do workforce members update the virus protection software when it is routed to them?						
		Do you prohibit workforce members from downloading software they brought in from elsewhere? (digital family photos, games, books, music, etc.)						
Log-in Monitoring	Addressable							
		Does the Security Official regularly monitor audit logs?						
		Is the Security Official notified of unsuccessful log-ins?						
		Do workforce members know what to do if they cannot access the system?						

(continued)

Sample HIPAA Security Risk Assessment
For a Small Physician Practice

Implementation Specification	R/A	Sample Risk Assessment Question	Risk			Policy		Assigned to
			Risk for Us	Could be a Risk	Not a Risk	Policy in Place	Need Policy	
Password Management	Addressable							
		Have you established procedures for creating, changing, and safeguarding passwords?						
		Are sanctions in place if workforce members share passwords?						
		Do workforce members know what to do if they forget a password?						
		Are you providing password management reminders?						
Security Incident Procedures 164.308(a)(6) **Team: Security Official, Practice Management Vendor**								
Response and Reporting	Required							
		Do you know if your security system has ever been breached?						
		Have you prioritized what must be restored in the event of a system disruption?						
		Have you developed a list of persons and entities to contact in the event of a security incident?						
		Do you require workforce members to tell you immediately if they suspect a compromise to your system?						

Contingency Plan 164.308(a)(7)
Team: Security Official, Privacy Official, Physician

		Have you made a list of possible security incidents?							
		Do you document all security incidents and their outcomes?							
Data Backup Plan	Required								
		Does your practice back up its electronic data?							
		Do you store the backup data at the physician practice's location?							
		Do you know whom to call to restore data?							
Disaster Recovery Plan	Required								
		Do you have a procedure to restore any loss of data?							
		Do you have a list of critical hardware, software, and workforce members?							
Emergency Mode Operation Plan	Required								
		If you are required to operate in emergency mode, do you have procedures to enable you to continue critical business processes to protect the security of ePHI?							
		Do you have a plan to temporarily relocate if you lose access to your physical location?							
		Would ePHI be safeguarded in this temporary location?							
		Are formal agreements in place for such a relocation?							
		Have you trained staff on your contingency plan?							
		Is there a contingency plan coordinator?							

(continued)

Sample HIPAA Security Risk Assessment
For a Small Physician Practice

Implementation Specification	R/A	Sample Risk Assessment Question	Risk			Policy		Assigned to
			Risk for Us	Could be a Risk	Not a Risk	Policy in Place	Need Policy	
		Do you have an emergency call list?						
		Have you identified situations in which your contingency plan must be activated?						
		Is there a plan to restore systems to your normal operations?						
Testing and Revision Procedures	Addressable							
		Have you tested your contingency plan?						
Applications and Data Criticality Analysis	Addressable							
		Do you have a plan to restore your business activities, beginning with what is most critical to your practice?						
Evaluation 164.308(a)(8) **Team: Security Official, Privacy Official, Physician**								
Evaluation	Required							
		Do you perform periodic HIPAA Security evaluations?						

	Do you perform these evaluations in response to environmental and operations changes affecting the security of your ePHI, to determine whether your security policies and procedures meet HIPAA Security requirements?				
	Do you perform both technical and nontechnical evaluations?				
	Has your Security Official determined acceptable levels of risk in its business operations and mitigation strategies?				
	Do you have a plan to evaluate your systems at least annually, or at any time a risk warrants a review?				

Business Associate Contracts and Other Arrangements 164.308(b)(1)

Team: Security Official, Privacy Official, Physician

	Required				
Written Business Associate Contract					
	Are all necessary Business Associate Agreements in place? Are they HIPAA and HITECH compliant?				
	Are there new organizations or IT vendors that require a Business Associate Agreement?				

(continued)

Sample HIPAA Security Risk Assessment
For a Small Physician Practice

Physical Safeguards

Facility Access Controls 164.310(a)(1)

Team: Security Official, Privacy Official, Physician

Implementation Specification	R/A	Sample Risk Assessment Question	Risk			Policy		Assigned to
			Risk for Us	Could be a Risk	Not a Risk	Policy in Place	Need Policy	
Contingency Operations	Addressable							
		Do you know who needs access to the facility in the event of a disaster?						
		Do you have a backup plan for access, including who has authority to access the facility in a disaster?						
Facility Security Plan	Addressable							
		Do you have an inventory of facilities and equipment therein?						
		How do you safeguard your facility and equipment from unauthorized physical access, tampering, and theft?						
		Is there a contingency plan in place?						
Access Control and Validation Procedures	Addressable							

		Do you have procedures in place to control physical access to your facility and areas within your facility where ePHI could be accessed?						
		Do you validate a person's authority to access software programs for testing and revision?						
		Is there a history or risk of break-ins that requires monitoring equipment?						
		If monitoring or surveillance equipment generates records or footage, how is it reviewed, handled, and disposed of?						
		If you use a security contractor for surveillance purposes, do you have an up-to-date Business Associate Agreement in place with the contractor?						
Maintenance Records	Addressable							
		Have you repaired or modified any physical components of your facility related to security, such as doors, locks, walls, or hardware, or do you expect to do so?						
		Do you have a system to document all such repairs and modifications?						

Workstation Use 164.310(b)
Team: Security Official, Privacy Official, Physician

Workstation Use	Required	Have you documented how workstations are to be used in the physician practice?						
		Are there wireless tools used as workstations?						
		Can unauthorized persons view content of workstations?						

(continued)

Sample HIPAA Security Risk Assessment
For a Small Physician Practice

Implementation Specification	R/A	Sample Risk Assessment Question	Risk			Policy		Assigned to
			Risk for Us	Could be a Risk	Not a Risk	Policy in Place	Need Policy	
Workstation Security 164.310(c)								
Team: Security Official, Privacy Official, Physician, IT Vendor								
Workstation Security	Required							
		Is access to ePHI restricted to authorized users?						
		Is there a log-off policy before leaving computers unattended?						
		Is there a policy that controls Internet access while working with ePHI?						
Device and Media Controls 164.310(d)(1)								
Team: Security Official, Privacy Official, Physician, IT Vendor								
Disposal	Required	Do you destroy data on hard drives and file servers before disposing the hardware?						
Media Re-use	Required	Are workforce members trained as to the security risks of re-using hardware and software that contain ePHI?						
		Do you have a procedure for removing ePHI from electronic media before it can be re-used?						
Accountability	Addressable	Do you document the movement of hardware and electronic media and who is responsible for each item?						

		Do you periodically check the inventory to ensure computers are where they are supposed to be?					
		Do you document where they've been moved?					
		Is the inventory list part of your disaster recovery files? Is it stored in a disaster-proof manner, i.e., offsite and (preferably) electronically?					
Data Backup and Storage	Addressable						
		Do you regularly back up data on hardware and software and maintain backup files off site?					
		Do you back up ePHI before equipment is moved?					
		Have staff members been trained on backup policies?					

Technical Safeguards

Access Control 164.312(a)(1)

Team: Security Official, Privacy Official, Physician, IT Vendor

Unique User Identification	Required						
		Has the Security Official assigned a unique user identity to each member of the workforce?					
		Are passwords unique to each individual and not shared?					
		Is there a sanction policy on sharing passwords?					
		Do workforce members have access to the minimum ePHI necessary to perform their job responsibilities?					

(continued)

Sample HIPAA Security Risk Assessment
For a Small Physician Practice

Implementation Specification	R/A	Sample Risk Assessment Question	Risk			Policy		Assigned to
			Risk for Us	Could be a Risk	Not a Risk	Policy in Place	Need Policy	
		Do you participate in ePrescribing? If so, does the system validate your electronic signature? Are physicians the only providers allowed to ePrescribe in your practice?						
Emergency Access Procedure	Required							
		Does the Security Official have a unique user ID that is used only in emergencies?						
		Is there a process to notify another leader in the practice when the emergency ID is used?						
Automatic Logoff	Addressable	Do your computers automatically log off after a specific period of inactivity?						
		Is there a shorter log off period for computers in high traffic areas?						
Encryption and Decryption	Addressable							
		Do you send e-mail containing ePHI to patients?						
		Is the e-mail sent over an open network such as AOL, Yahoo!, EarthLink, or Comcast?						
		Do you have a mechanism in place to encrypt and decrypt ePHI?						

Audit Controls 164.312(b)

Team: Security Official, IT Vendor

Audit Controls	Required						
	Is there a procedure in place to monitor and audit workforce members with access to ePHI and their activity with respect to ePHI?						
	Is one person responsible for conducting audit processes and reporting results?						
	Has your IT vendor explained how to conduct audits?						

Integrity 164.312(c)(1)

Team: Security Official, Privacy Official, Physician, IT Vendor

Mechanism to Authenticate ePHI	Addressable						
	Are users required to authenticate themselves when logging on to the system?						
	Is there a feature that locks out users after a specific number of failed log-in attempts?						
	Is data transmitted through standard network protocols?						
	Have you identified sources that would jeopardize the integrity of ePHI (vandalism, hackers, system failures, viruses)?						
	Is there an electronic mechanism to corroborate that ePHI has not been altered or destroyed in an unauthorized manner?						

(continued)

Sample HIPAA Security Risk Assessment
For a Small Physician Practice

Implementation Specification	R/A	Sample Risk Assessment Question	Risk			Policy		Assigned to
			Risk for Us	Could be a Risk	Not a Risk	Policy in Place	Need Policy	
		Person or Entity Authentication 164.312(d) **Team: Security Official, Privacy Official, Physician, IT Vendor**						
Person or Entity Authentication	Required							
		Does your system require users to identify themselves using a password and user name?						
		Does the system allow you to conduct audit trails on users?						
		Transmission Security 164.312(e)(1) **Team: Security Official, IT Vendor**						
Integrity Controls	Addressable							
		Does the software allow you to track and audit users who transmit and alter ePHI?						
		Is there an auditing process in place?						
		Does the IT vendor ensure that information is not altered in transmission?						
Encryption	Addressable							
		Does your practice use a mechanism (secure network) to encrypt e-mail or other ePHI?						
		Does your practice send ePHI via handhelds or wireless laptops?						
		Do workforce members know how to respond to e-mails containing ePHI?						

Breach Notification Rule

Notification in the Case of Breach of Unsecured Protected Health Information ("PHI") 45 CFR Part 164 Subpart D

Team: Security Official, Physician, Workforce Members

Are paper charts or portable computers containing PHI ever taken out of the practice? This includes portable computers, back-up tapes, smart phones, paper charts.							
Are electronic devices encrypted? Do you periodically check electronic equipment to ensure encryption safeguards have not been disabled?							
How does your practice "secure" non-electronic PHI ("secure" has a specific meaning under the Breach Notification Rule) and protect oral PHI?							
Is your workforce trained to immediately report suspected breaches of PHI?							
Does your practice have a procedure in place to conduct a risk analysis of any suspected breaches of PHI?							
Have you asked your Business Associates what they are doing to comply with the Breach Notification Rule?							
Have you updated your Business Associate Agreements to require your Business Associates to notify you promptly if they discover a breach of PHI and to provide you with all of the appropriate information regarding the breach?							

(continued)

Sample HIPAA Security Risk Assessment
For a Small Physician Practice

Implementation Specification	R/A	Sample Risk Assessment Question	Risk			Policy		
			Risk for Us	Could be a Risk	Not a Risk	Policy in Place	Need Policy	Assigned to

Reproduction of this material by physicians and their staff is permitted. Any other use, duplication or distribution by any other party requires the prior written approval of the American Physician Association. This material is educational only, does not constitute legal advice, and covers only federal, not state, law. Changes in applicable laws or regulations may require revision. Physicians should contact their personal attorneys for legal advice pertaining to HIPAA compliance, the HITECH Act, and the U.S. Department of Health and Human Services rules and regulations.

Physicians EHR, Inc. l 2500 Regency Parkway l Cary, North Carolina 27518
Tel: (919) 859-9907 l Fax: (919) 573-0430 l Toll Free: (800) 906-6347

Appendix C

CHEMICALS OF INTEREST FOR SECURITY

Chemicals of Interest	Synonym	Theft		Sabotage		Security Issues						
		Min. Concentration (%)	Screening Threshold Quantities (in pounds unless otherwise noted)	Min. Concentration (%)	Screening Threshold Quantities (in pounds)	Release–Toxic	Release–Flammables	Release–Explosives	Theft–CW/CWP	Theft–W/ME	Theft–EXP/IEDP	Sabotage/Contamination
Acetaldehyde							X					
Acetone cyanohydrin, stabilized				ACG	APA							X
Acetyl bromide				ACG	APA							X
Acetyl chloride				ACG	APA							X
Acetyl iodide				ACG	APA							X
Acetylene	Ethyne						X					
Acrolein	2-Propenal; Acrylaldehyde					X						
Acrylonitrile	2-Propenenitrile						X					
Acrylyl chloride	2-Propenoyl chloride						X					
Allyl alcohol	2-Propen-1-ol					X						
Allylamine	2-Propen-1-amine						X					
Allyltrichlorosilane, stabilized				ACG	APA							X
Aluminum (powder)		ACG	100								X	
Aluminum bromide, anhydrous				ACG	APA							X
Aluminum chloride, anhydrous				ACG	APA							X
Aluminum phosphide				ACG	APA							X
Ammonia (anhydrous)						X						

Ammonia (conc. 20% or greater)						X					
Ammonium nitrate, [with more than 0.2% combustible substances, including any organic substance calculated as carbon, to the exclusion of any other added substance]		ACG	400				X			X	
Ammonium nitrate, solid [nitrogen concentration of 23% nitrogen or greater]		33	2000							X	
Ammonium perchlorate		ACG	400				X			X	
Ammonium picrate		ACG	400				X			X	
Amyltrichlorosilane				ACG	APA						X
Antimony pentafluoride				ACG	APA						X
Arsenic trichloride	Arsenous trichloride	30	2.2			X		X			
Arsine		0.67	15			X			X		
Barium azide		ACG	400				X			X	
1,4-Bis(2-chloroethylthio)-nbutane		CUM 100g					X				
Bis(2-chloroethylthio)methane		CUM 100g					X				
Bis(2-chloroethylthiomethyl) ether		CUM 100g					X				
1,5-Bis(2-chloroethylthio)-npentane		CUM 100g					X				
1,3-Bis(2-chloroethylthio)-npropane		CUM 100g					X				

Chemicals of Interest	Synonym	Theft Min. Concentration (%)	Theft Screening Threshold Quantities (in pounds unless otherwise noted)	Sabotage Min. Concentration (%)	Sabotage Screening Threshold Quantities (in pounds)	Release–Toxic	Release–Flammables	Release–Explosives	Theft–CW/CWP	Theft–WME	Theft–EXP/IEDP	Sabotage/Contamination
Boron tribromide		12.67	45	ACG	APA					X		X
Boron trichloride	Borane, trichloro	84.7	45			X				X		X
Boron trifluoride	Borane, trifluoro	26.87	45			X				X		
Boron trifluoride compound with methyl ether (1:1)	Boron, trifluoro [oxybis (methane)], T-4					X						
Bromine						X						
Bromine chloride		9.67	45							X		
Bromine pentafluoride				ACG	APA							X
Bromotrifluorethylene	Ethene, bromotrifluoro-						X					
1,3-Butadiene							X					
Butane							X					
Butene							X					
1-Butene							X					
2-Butene							X					
2-Butene-cis							X					
2-Butene-trans	2-Butene, (E)						X					
Butyltrichlorosilane				ACG	APA							X
Calcium hydrosulfite	Calcium dithionite			ACG	APA							X
Calcium phosphide				ACG	APA							X
Carbon disulfide						X						
Carbon oxysulfide	Carbon oxide sulfide (COS);carbonyl sulfide						X					

Name	Synonym									
Carbonyl fluoride		12	45							X	
Carbonyl sulfide		56.67	500							X	
Chlorine		9.77	500			X				X	X
Chlorine dioxide	Chlorine oxide, (ClO2)			ACG	APA	X					
Chlorine monoxide	Chlorine oxide						X				
Chlorine pentafluoride		4.07	15							X	
Chlorine trifluoride		9.97	45							X	
Chloroacetyl chloride				ACG	APA						X
2-Chloroethylchloromethylsulfide		CUM 100g						X			
Chloroform	Methane, trichloro-					X					
Chloromethyl methylether	Methane, chloromethoxy-					X					
1-Chloropropylene	1-Propene, 1-chloro-						X				
2-Chloropropylene	1-Propene, 2-chloro-						X				
Chlorosarin	o-Isopropyl methylphosphonochloridate	CUM 100g						X			
Chlorosoman	o-Pinacolyl methylphosphonochloridate	CUM 100g						X			
Chlorosulfonic acid				ACG	APA						X
Chromium oxychloride				ACG	APA						X
Crotonaldehyde	2-Butenal						X				
Crotonaldehyde, (E)-	2-Butenal, (E)-	11.67	45				X				
Cyanogen	Ethanedinitrile	2.67	15				X			X	
Cyanogen chloride						X				X	
Cyclohexylamine	Cyclohexanamine					X					
Cyclohexyltrichlorosilane				ACG	APA						X
Cyclopropane						X					

Chemicals of Interest	Synonym	Theft Min. Concentration (%)	Theft Screening Threshold Quantities (in pounds unless otherwise noted)	Sabotage Min. Concentration (%)	Sabotage Screening Threshold Quantities (in pounds)	Release–Toxic	Release–Flammables	Release–Explosives	Theft–CW/CWP	Theft–WME	Theft–EXP/IEDP	Sabotage/Contamination
DF	Methyl phosphonyl difluoride	CUM 100g						X				
Diazodinitrophenol		ACG	400					X			X	
Diborane		2.67	15			X				X		
Dichlorosilane	Silane, dichloro-	10.47	45				X			X		
N,N-(2-diethylamino)ethanethiol		30	2.2						X			
Diethyldichlorosilane				ACG	APA							X
Diethyleneglycol dinitrate		ACG	400					X			X	
Diethyl methylphosphonite		30	2.2						X			
N,N-Diethyl phosphoramidic dichloride		30	2.2						X			
N,N-(2-diisopropylamino)ethanethiol N,N-diisopropyl-(beta)-aminoethane thiol		30	2.2						X			
Difluoroethane	Ethane, 1,1-difluoro-						X					
N,N-Diisopropyl phosphoramidic dichloride		30	2.2						X			
1,1-Dimethylhydrazine	Hydrazine, 1,1-dimethyl-						X					
Dimethylamine	Methanamine, N-methyl-						X					
N,N-(2-dimethylamino)ethanethiol		30	2.2						X			

Chemical	Synonym											
Dimethyldichlorosilane	Silane, dichlorodimethyl-											X
N,N-Dimethyl phosphoramidic dichloride		30	2.2						X			
Dimethyl-phosphoramido-dichloridate												
2,2-Dimethylpropane	Propane, 2,2-dimethyl-						X	X				
Dingu	Dinitroglycoluril	ACG	400					X			X	
Dinitrogen tetroxide		3.8	15							X		
Dinitrophenol		ACG	400					X			X	
Dinitroresorcinol		ACG	400					X			X	
Diphenyldichlorosilane				ACG	APA							X
Dipicryl sulfide		ACG	400					X			X	
N,N-(2-dipropylamino) ethanethiol		30	2.2						X			
N,N-Dipropyl phosphoramidic dichloride		30	2.2						X			
Dodecyltrichlorosilane				ACG	APA							X
Epichlorohydrin	Oxirane, (chloromethyl)-					X						
Ethane							X					
Ethyl acetylene	1-Butyne						X					
Ethyl chloride	Ethane, chloro-						X					
Ethyl ether	Ethane, 1,1'-oxybis-						X					
Ethyl mercaptan	Ethanethiol						X					
Ethyl nitrite	Nitrous acid, ethyl ester						X					
Ethyl phosphonyl difluoride		CUM 100g						X				
Ethylamine	Ethanamine						X					
Ethyldiethanolamine		80	220						X			

		Theft		Sabotage		Security Issues						
Chemicals of Interest	Synonym	Min. Concentration (%)	Screening Threshold Quantities (in pounds unless otherwise noted)	Min. Concentration (%)	Screening Threshold Quantities (in pounds)	Release–Toxic	Release–Flammables	Release–Explosives	Theft–CW/CWP	Theft–WME	Theft–EXP/IEDP	Sabotage/Contamination
Ethylene	Ethene						X					
Ethylene oxide	Oxirane						X					
Ethylenediamine	1,2-Ethanediamine					X						
Ethyleneimine	Aziridine						X					
Ethylphosphonothioic dichloride		30	2.2						X			
Ethyltrichlorosilane				ACG	APA							X
Fluorine		6.17	15			X				X		
Fluorosulfonic acid				ACG	APA							X
Formaldehyde (solution)						X						
Furan							X					
Germanium tetrafluoride		2.11	15						X	X		
Guanyl nitrosaminoguanylidene hydrazine		ACG	400					X			X	
Hexaethyl tetraphosphate and compressed gas mixtures		33.37	500							X		
Hexafluoroacetone		15.67	45							X		
Hexanitrostilbene		ACG	400					X			X	
Hexolite	Hexotol	ACG	400					X			X	
Hexyltrichlorosilane				ACG	APA							X

Chemical	Synonym									
HMX	Cyclotetramethyl-enetetranitramine	ACG	400							X
HN1 (nitrogen mustard-1)	Bis(2-chloroethyl) ethylamine	CUM 100g						X		
HN2 (nitrogen mustard-2)	Bis(2-chloroethyl) methylamine	CUM 100g						X		
HN3 (nitrogen mustard-3)	Tris(2-chloroethyl)amine	CUM 100g						X		
Hydrazine						X				
Hydrochloric acid (conc. 37% or greater)					X					
Hydrocyanic acid					X			X		
Hydrofluoric acid (conc. 50% or greater)					X			X		
Hydrogen						X				
Hydrogen bromide (anhydrous)		95.33	500		X					
Hydrogen chloride (anhydrous)		ACG	500		X					
Hydrogen cyanide	Hydrocyanic acid	4.67	15		X			X		
Hydrogen fluoride (anhydrous)		42.53	45		X			X		
Hydrogen peroxide (concentration of at least 35%)		35	400	APA					X	
Hydrogen selenide		0.07	15		X			X		
Hydrogen sulfide		23.73	45		X			X		
Iodine pentafluoride				ACG						X
Iron, pentacarbonyl-	Iron carbonyl (Fe (CO)5), (TB5-11)-				X	X				
Isobutane	Propane, 2-methyl									
Isobutyronitrile	Propanenitrile, 2-methyl-				X					

Chemicals of Interest	Synonym	Theft		Sabotage		Security Issues						
		Min. Concentration (%)	Screening Threshold Quantities (in pounds unless otherwise noted)	Min. Concentration (%)	Screening Threshold Quantities (in pounds)	Release–Toxic	Release–Flammables	Release–Explosives	Theft–CW/CWP	Theft–WME	Theft–EXP/IEDP	Sabotage/Contamination
Isopentane	Butane, 2-methyl-						X					
Isoprene	1,3-Butadiene, 2-methyl-						X					
Isopropyl chloride	Propane, 2-chloro-						X					
Isopropyl chloroformate	Carbonochloridic acid, 1-methylethyl ester					X						
Isopropylamine	2-Propanamine						X					
Isopropylphosphonothioic dichloride		30	2.2						X			
Isopropylphosphonyl difluoride		CUM 100g						X				
Lead azide		ACG	400					X			X	
Lead styphnate	Lead trinitroresorcinate	ACG	400					X			X	
Lewisite 1	2-Chlorovinyldi-chloroarsine	CUM 100g						X				
Lewisite 2	Bis(2-chlorovinyl) chloroarsine	CUM 100g						X				
Lewisite 3	Tris(2-chlorovinyl)arsine	CUM 100g						X				
Lithium amide				ACG	APA							X
Magnesium (powder)		ACG	100								X	
Magnesium diamide				ACG	APA							X
Magnesium phosphide				ACG	APA							X
MDEA	Methyldiethanolamine	80	220						X			
Mercury fulminate		ACG	400					X			X	

Chemical Name	Synonym										
Methacrylonitrile	2-Propenenitrile,2-methyl-					X					
Methane							X				
2-Methyl-1-butene							X				
3-Methyl-1-butene							X				
Methyl chloride	Methane, chloro-						X				
Methyl chloroformate	Carbonochloridic acid, methyl ester						X				
Methyl ether	Methane, oxybis-						X				
Methyl formate	Formic acid Methyl ester						X				
Methyl hydrazine	Hydrazine, methyl-					X					
Methyl isocyanate	Methane, isocyanato-					X					
Methyl mercaptan	Methanethiol	45	500			X	X		X		
Methyl thiocyanate	Thiocyanic acid, methyl ester					X					
Methylamine	Methanamine						X		X		
Methylchlorosilane		20	45								
Methyldichlorosilane				ACG	APA						X
Methylphenyl-dichlorosilane				ACG	APA						X
Methylphosphonothioic dichloride		30	2.2					X			
Methyltrichlorosilane	Silane, trichloromethyl-			ACG	APA		X				X
Sulfur mustard (Mustard gas(H))	Bis(2-chloroethyl)sulfide	CUM 100g									
O-Mustard (T)	Bis(2-chloroethylthioethyl) ether	CUM 100g									
Nickel Carbonyl							X				
Nitric acid		68	400			X				X	

Chemicals of Interest	Synonym	Theft Min. Concentration (%)	Theft Screening Threshold Quantities (in pounds unless otherwise noted)	Sabotage Min. Concentration (%)	Sabotage Screening Threshold Quantities (in pounds)	Release–Toxic	Release–Flammables	Release–Explosives	Theft–CW/CWP	Theft–WME	Theft–EXP/IEDP	Sabotage/Contamination
Nitric oxide	Nitrogen oxide (NO)	3.83	15			X						
Nitrobenzene		ACG	100								X	
5-Nitrobenzotriazol		ACG	400					X			X	
Nitrocellulose		ACG	400					X			X	
Nitrogen mustard hydrochloride	Bis(2-chloroethyl) methylamine hydrochloride	30	2.2						X			
Nitrogen trioxide		3.83	15							X		
Nitroglycerine		ACG	400					X			X	
Nitromannite	Mannitol hexanitrate, wetted	ACG	400					X			X	
Nitromethane		ACG	400								X	
Nitrostarch		ACG	400					X			X	
Nitrosyl chloride		1.17	15							X		
Nitrotriazolone		ACG	400					X			X	
Nonyltrichlorosilane				ACG	APA							X
Octadecyltrichlorosilane				ACG	APA							X
Octolite		ACG	400					X			X	
Octonal		ACG	400					X			X	
Oleum (Fuming Sulfuric acid)	Sulfuric acid, mixture with sulfur trioxide					X						
Oxygen difluoride		0.09	15							X		
1,3-Pentadiene							X					

Name	Synonym									
Pentane						X				
1- Pentene						X				
2-Pentene, (E)-						X				
2-Pentene, (Z)-						X				
Pentolite		ACG	400							X
Peracetic acid	Ethaneperoxic acid						X	X		
Perchloromethylmercaptan	Methanesulfenyl chloride, trichloro-						X			
Perchloryl fluoride		25.67	45					X	X	
PETN	Pentaerythritol tetranitrate	ACG	400					X	X	X
Phenyltrichlorosilane				ACG	APA					X
Phosgene	Carbonic dichloride;carbonyl dichloride	0.17	15				X	X	X	
Phosphine		0.67	15					X	X	
Phosphorus		ACG	400					X		
Phosphorus oxychloride	Phosphoryl chloride	80	220	ACG	APA		X	X		X
Phosphorus pentabromide				ACG	APA					X
Phosphorus pentachloride				ACG	APA					X
Phosphorus pentasulfide				ACG	APA					X
Phosphorus trichloride		3.48	45	ACG	APA		X		X	X
Piperidine						X				
Potassium chlorate		ACG	400						X	
Potassium cyanide				ACG	APA					X
Potassium nitrate		ACG	400						X	
Potassium perchlorate		ACG	400						X	
Potassium permanganate		ACG	400						X	
Potassium phosphide		ACG	400	ACG	APA					X

Chemicals of Interest	Synonym	Theft Min. Concentration (%)	Theft Screening Threshold Quantities (in pounds unless otherwise noted)	Sabotage Min. Concentration (%)	Sabotage Screening Threshold Quantities (in pounds)	Release– Toxic	Release– Flammables	Release– Explosives	Theft– CW/ CWP	Theft– WME	Theft– EXP/ IEDP	Sabotage/ Contamination
Propadiene	1,2-Propadiene						X					
Propane							X					
Propionitrile	Propanenitrile					X						
Propyl chloroformate	Carbonochloridic acid, propylester						X					
Propylene [1-Propene]							X					
Propylene oxide	Oxirane, methyl-						X					
Propyleneimine	Aziridine, 2-methyl-					X						
Propylphosphonothioic dichloride		30	2.2						X			
Propylphosphonyl difluoride		CUM 100g						X				
Propyltrichlorosilane				ACG	APA							X
Propyne	1-Propyne						X	X				
QL	o-Ethyl-o-2-diisopropylaminoethyl methylphosphonite	CUM 100g						X				
RDX	Cyclotrimethylene-trinitramine	ACG	400					X			X	
RDX and HMX mixtures		ACG	400					X			X	
Selenium hexafluoride		1.67	15							X		
Sesquimustard	1,2-Bis(2-chloroethylthio)ethane	CUM 100g						X				

Name	Synonym										
Silane							X				
Silicon tetrachloride		15		ACG	APA						X
Silicon tetrafluoride		ACG	45						X	X	
Sodium azide		ACG	400						X		
Sodium chlorate		ACG	400						X	X	
Sodium cyanide				ACG	APA						X
Sodium hydrosulfite	Sodium dithionite			ACG	APA						X
Sodium nitrate		ACG	400	ACG						X	
Sodium phosphide				ACG	APA						X
Soman	o-Pinacolyl methylphosphonofluoridate	CUM 100g				X					
Stibine		0.67	15	ACG					X		X
Strontium phosphide					APA						
Sulfur dioxide (anhydrous)		84	500					X	X		
Sulfur tetrafluoride	Sulfurfluoride(SF4),(T-4)-	1.33	15					X	X		
Sulfur trioxide								X			
Sulfuryl chloride				ACG	APA						X
Tabun	o-Ethyl-N,N-dimethyl-phosphoramido-cyanidate	CUM 100g				X					
Tellurium hexafluoride		0.83	15						X		
Tetrafluoroethylene	Ethene, tetrafluoro-						X				
Tetramethylsilane	Silane, tetramethyl-						X				
Tetranitroaniline		ACG	400			X				X	
Tetranitromethane	Methane, tetranitro-					X	X				
Tetrazene	Guanyl nitrosaminoguanyltetrazene	ACG	400			X				X	

Chemicals of Interest	Synonym	Theft – Min. Concentration (%)	Theft – Screening Threshold Quantities (in pounds unless otherwise noted)	Sabotage – Min. Concentration (%)	Sabotage – Screening Threshold Quantities (in pounds)	Release–Toxic	Release–Flammables	Release–Explosives	Theft–CW/CWP	Theft–WME	Theft–EXP/IEDP	Sabotage/Contamination
1H-Tetrazole		ACG	400					X			X	
Thiodiglycol	Bis(2-hydroxyethyl) sulfide	30	2.2						X			
Thionyl chloride				ACG	APA							X
Titanium tetrachloride	Titanium chloride (TiCl4) (T-4)-	13.33	45	ACG	APA	X				X		X
TNT	Trinitrotoluene	ACG	400					X			X	
Torpex	Hexotonal	ACG	400					X			X	
Trichlorosilane	Silane, trichloro-	80	220	ACG	APA		X					X
Triethanolamine		80	220						X			
Triethanolamine hydrochloride		80	220						X			
Triethyl phosphate		80	220						X			
Trifluoroacetyl chloride		6.93	45							X		
Trifluorochloroethylene	Ethene, chlorotrifluoro	66.67	500				X			X		
Trimethylamine,N,N-dimethyl-	Methanamine,N,N-dimethyl-						X					
Trimethylchlorosilane	Silane, chlorotrimethyl-			ACG	APA		X					
Trimethyl phosphate		80	220						X			X
Trinitroaniline		ACG	400					X			X	
Trinitroanisole		ACG	400					X			X	
Trinitrobenzenesulfonic acid		ACG	400					X			X	
Trinitrobenzoic acid		ACG	400					X			X	

Name	Synonym									
Trinitrochlorobenzene		ACG	400				X		X	
Trinitrofluorenone		ACG	400				X		X	
Trinitro-meta-cresol		ACG	400				X		X	
Trinitronaphthalene		ACG	400				X		X	
Trinitrophenetole		ACG	400				X		X	
Trinitrophenol		ACG	400				X		X	
Trinitroresorcinol		ACG	400				X		X	
Tritonal		ACG	400				X		X	
Tungsten hexafluoride		7.1	45					X		
Vinyl acetate monomer	Acetic acid ethenyl ester					X				
Vinyl acetylene	1-Buten-3-yne					X				
Vinyl chloride	Ethene, chloro-					X				
Vinyl ethyl ether	Ethene, ethoxy-					X				
Vinyl fluoride	Ethene, fluoro-					X				
Vinyl methyl ether	Ethene, methoxy-					X				
Vinylidene chloride	Ethene, 1,1-dichloro-					X				
Vinylidene fluoride	Ethene, 1,1-difluoro-					X				
Vinyltrichlorosilane				ACG	APA					X
VX	o-Ethyl-S-2-diisopropylaminoethyl methyl phosphonothiolate	CUM 100g					X			
Zinc hydrosulfite	Zinc dithionite			ACG	APA					X

Appendix D

GLOSSARY

Accidental Hazard: Source of harm or difficulty created by negligence, error, or unintended failure.

Accountability: The security goal that generates the requirement for actions of an entity to be traced uniquely to that entity.

Adversary: Individual, group, organization, or government that conducts or has the intent to conduct detrimental activities.

Asset: Person, structure, facility, information, material, or process that has value.

Assurance: Confidence that the other four security goals (integrity, availability, confidentiality, and accountability) have been adequately met.

Attack Method: Manner and means, including the weapon and delivery method, an adversary may use to cause harm on a target.

Attack Path: Steps that an adversary takes or may take to plan, prepare for, and execute an attack.

Availability: The security goal that generates the requirement for protection against—intentional or accidental attempts to (1) perform unauthorized deletion of data or (2) otherwise cause a denial of service or data.

Capability: Means to accomplish a mission, function, or objective.

Confidentiality: The security goal that generates the requirement for protection from intentional or accidental attempts to perform unauthorized data reads.

Consequence: Effect of an event, incident, or occurrence.

Consequence Assessment: Process of identifying or evaluating the potential or actual effects of an event, incident, or occurrence.

Countermeasure: Action, measure, or device that reduces an identified risk.

Deterrent: Measure that discourages an action or prevents an occurrence by instilling fear, doubt, or anxiety.

Due Care: A duty to provide for information security to ensure that the type of control, the cost of control, and the deployment of control are appropriate for the system being managed.

Economic Consequence: Effect of an incident, event, or occurrence on the value of property or on the production, trade, distribution, or use of income, wealth, or commodities.

Evaluation: Process of examining, measuring, and/or judging how well an entity, procedure, or action has met or is meeting stated objectives.

Function: Service, process, capability, or operation performed by an asset, system, network, or organization.

Hazard: Natural or manmade source or cause of harm or difficulty.

Human Consequence: Effect of an incident, event, or occurrence that results in injury, illness, or loss of life.

Implementation: Act of putting a procedure or course of action into effect to support goals or achieve objectives.

Incident: Occurrence caused by either human action or natural phenomena that may cause harm and that may require action.

Integrated Risk Management: Incorporation and coordination of strategy, capability, and governance to enable risk-informed decision making.

Integrity: The security goal that generates the requirement for protection against either intentional or accidental attempts to violate data integrity.

IT Security Goals: The five security goals are integrity, availability, confidentiality, accountability, and assurance.

Intent: Determination to achieve an objective.

Intentional Hazard: Source of harm, duress, or difficulty created by a deliberate action or a planned course of action.

Likelihood: Estimate of the potential of an incident or event's occurrence.

Mission Consequence: Effect of an incident, event, operation, or occurrence on the ability of an organization or group to meet a strategic objective or perform a function.

Model: Approximation, representation, or idealization of selected aspects of the structure, behavior, operation, or other characteristics of a real-world process, concept, or system.

Natural Hazard: Source of harm or difficulty created by a meteorological, environmental, or geological phenomenon or combination of phenomena.

Network: Group of components that share information or interact with each other in order to perform a function.

Probabilistic Risk Assessment: Type of quantitative risk assessment that considers possible combinations of occurrences with associated consequences, each with an associated probability or probability distribution.

Probability (Mathematical): Likelihood that is expressed as a number between 0 and 1, where 0 indicates that the occurrence is impossible and 1 indicates definite knowledge that the occurrence has happened or will happen, where the ratios between numbers reflect and maintain quantitative relationships.

Psychological Consequence: Effect of an incident, event, or occurrence on the mental or emotional state of individuals or groups resulting in a change in perception and/or behavior.

Qualitative Risk Assessment Methodology: Set of methods, principles, or rules for assessing risk based on nonnumerical categories or levels.

Quantitative Risk Assessment Methodology: Set of methods, principles, or rules for assessing risks based on the use of numbers where the meanings and proportionality of values are maintained inside and outside the context of the assessment.

Redundancy: Additional or alternative systems, subsystems, assets, or processes that maintain a degree of overall functionality in case of loss or failure of another system, subsystem, asset, or process.

Residual Risk: Risk that remains after risk management measures have been implemented.

Resilience: Ability to resist, absorb, recover from, or successfully adapt to adversity or a change in conditions.

Return on Investment (Risk): Calculation of the value of risk reduction measures in the context of the cost of developing and implementing those measures.

Risk: Potential for an unwanted outcome resulting from an incident, event, or occurrence, as determined by its likelihood and the associated consequences.

Risk Acceptance: Explicit or implicit decision not to take an action that would affect all or part of a particular risk.

Risk Analysis: Systematic examination of the components and characteristics of risk.

Risk Assessment: Product or process that collects information and assigns values to risks for the purpose of informing priorities, developing or comparing courses of action, and informing decision making.

Risk Assessment Methodology: Set of methods, principles, or rules used to identify and assess risks and to form priorities, develop courses of action, and inform decision making.

Risk Assessment Tool: Activity, item, or program that contributes to determining and evaluating risks.

Risk Avoidance: Strategies or measures taken that effectively remove exposure to a risk.

Risk-Based Decision Making: Determination of a course of action predicated primarily on the assessment of risk and the expected impact of that course of action on that risk.

Risk Communication: Exchange of information with the goal of improving risk understanding, affecting risk perception, and/or equipping people or groups to act appropriately in response to an identified risk.

Risk Control: Deliberate action taken to reduce the potential for harm or maintain it at an acceptable level.

Risk Identification: Process of finding, recognizing, and describing potential risks.

Risk-Informed Decision Making: Determination of a course of action predicated on the assessment of risk, the expected impact of that course of action on that risk, as well as other relevant factors.

Risk Management: Process of identifying, analyzing, assessing, and communicating risk and accepting, avoiding, transferring, or controlling it to an acceptable level at an acceptable cost.

Risk Management Alternatives Development: Process of systematically examining risks to develop a range of options and their anticipated effects for decision makers.

Risk Management Cycle: Sequence of steps that are systematically taken and revisited to manage risk.

Risk Management Methodology: Set of methods, principles, or rules used to identify, analyze, assess, and communicate risk, and mitigate, accept, or control it to an acceptable level at an acceptable cost.

Risk Management Plan: Document that identifies risks and specifies the actions that have been chosen to manage those risks.

Risk Management Strategy: Course of action or actions to be taken in order to manage risks.

Risk Matrix: Tool for ranking and displaying components of risk in an array.

Risk Mitigation: Application of measure or measures to reduce the likelihood of an unwanted occurrence and/or its consequences.

Risk Mitigation Option: Measure, device, policy, or course of action taken with the intent of reducing risk.

Risk Perception: Subjective judgment about the characteristics and/or severity of risk.

Risk Profile: Description and/or depiction of risks to an asset, system, network, geographic area, or other entity.

Risk Reduction: Decrease in risk through risk avoidance, risk control, or risk transfer.

Risk Score: Numerical result of a semiquantitative risk assessment methodology.

Risk Tolerance: Degree to which an entity is willing to accept risk.

Risk Transfer: Action taken to manage risk that shifts some or all of the risk to another entity, asset, system, network, or geographic area.

Scenario (Risk): Hypothetical situation comprised of a hazard, an entity impacted by that hazard, and associated conditions including consequences when appropriate.

Semiquantitative Risk Assessment Methodology: Set of methods, principles, or rules to assess risk that uses bins, scales, or representative numbers whose values and meanings are not maintained in other contexts.

Sensitivity Analysis: Process to determine how outputs of a methodology differ in response to variation of the inputs or conditions.

Simulation: Model that behaves or operates like a given process, concept, or system when provided a set of controlled inputs.

Subject Matter Expert: Individual with in-depth knowledge in a specific area or field.

System: Any combination of facilities, equipment, personnel, procedures, and communications integrated for a specific purpose foundation for a useful transit system.

Target: Asset, network, system, or geographic area chosen by an adversary to be impacted by an attack.

Threat: Natural or manmade occurrence, individual, entity, or action that has or indicates the potential to harm life, information, operations, the environment, and/or property.

Threat Analysis: The examination of threat sources against system vulnerabilities to determine the threats for a particular system in a particular operational environment.

Threat Assessment: Process of identifying or evaluating entities, actions, or occurrences, whether natural or manmade, that have or indicate the potential to harm life, information, operations, and/or property sectors.

Threat Source: Either (1) intent and method targeted at the intentional exploitation of a vulnerability or (2) a situation and method that may accidentally trigger a vulnerability.

Uncertainty: Degree to which a calculated, estimated, or observed value may deviate from the true value.

Vulnerability: Physical feature or operational attribute that renders an entity open to exploitation or susceptible to a given hazard device.

Vulnerability Assessment: Process for identifying physical features or operational attributes that render an entity, asset, system, network, or geographic area susceptible or exposed to hazards.

Index